Group-target Tracking

Wen-dong Geng · Yuan-qin Wang
Zheng-hong Dong

Group-target Tracking

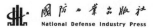
National Defense Industry Press

Wen-dong Geng
Equipment Academy
Beijing
China

Zheng-hong Dong
Equipment Academy
Beijing
China

Yuan-qin Wang
Equipment Academy
Beijing
China

Translated by Ge Geng and Fan Yang

ISBN 978-981-10-9470-5   ISBN 978-981-10-1888-6   (eBook)
DOI 10.1007/978-981-10-1888-6

Jointly published with National Defense Industry Press

Printed on acid-free paper

This Springer imprint is published by Springer Nature
The registered company is Springer Science+Business Media Singapore Pte Ltd.

Preface

As an important issue in military and civil researches, target tracking has got brilliant achievements not only in theory but also in practice. Meanwhile, we are expecting a continuous development in it. In terms of theory and technology, target tracking is classified into two categories—(a) Single-target tracking and multi-target tracking by single sensor based on Kalman filtering theory. (b) Single-target tracking and multi-target tracking by multi-sensor based on information fusion. In particular to the latter, the tracking to multi-target in high density is becoming a highlight and a challenge.

In some sense, multi-target tracking is actually an issue related to processing of multi-measurement data. For target tracking systems, however, multi-target naturally generates multi-measurement data, which are not necessarily resulted from multi-targets. For instance, an extended single target generates multi-measurement data. Meanwhile, high-resolution radar, multi-sensor, and multi-path effect could also be the resources. Such multi-measurement data are sometimes mixed with true and false information, which are even hard to select. For example, as false targets, interference, and other factors lead to mixture of true and false multi-measurement data, it is hard to distinguish the multi-measurement data produced by extended targets. The situation would be more complicated when there come the so-called indirect measurement data produced by multi-targets in high density with multi-path effect. Moreover, in the current target tracking system, target tracking and situation perception are separated at the physical layer. That is, the state estimation of every target is given first, and then a special information processing platform perceives the situation of multi-targets. Is it possible to find a new approach which integrates single-target and multi-target tracking while estimating the situation by connecting the physical layer, data layer, and perception layer without considering the quantity and density of targets?

Apparently, we cannot find such an approach unless we change our way of thinking. For this purpose, the book puts forward the concept of group-target tracking.

Group-target tracking is a multi-target centroid tracking method based on modern target tracking theories and formation tracking approach. It unifies group-target data association and situation association, as well as integrates group-target maneuver judgment, situation assessment, and combination/separation detection. Group-target tracking breaks through the restrictions on the quantity and density of targets as well as the restriction that requires one-to-one correspondence between real measurement data and relevant targets. It overcomes such problems as obscure data association for multi-target tracking in high density, inability to choose real measurement data from multi-returns of single target, and loss of useful information for target tracking during processing of indirect measurement data as false measurement data. As a result, it enables connection between the physical layer, data layer, and perception layer. Group-target tracking is a complete system capable of determining how a group-target is formed, how a group-target track initiates, how to make group-target association and track maintenance, and how to detect when new targets join in and old ones leave the group. It is also capable of providing prior information for capture of targets separated from the group and enabling group-target track termination. In fact, from a bionic perspective, group-target tracking is also a reverse situation perception process based on animals' vision process—first the whole picture, then each part, and finally each individual target.

Centered on group-target tracking, with consideration of perception issues such as situation association, situation fusion, and intention judgment, the book elaborates essential issues including formation of group-targets, track initiation, data association, combination/separation detection, and group-target track termination. The originality and novelty of the book lies in multiple aspects: (1) It proposes the group-target tracking concept featured by the unification of single-target tracking and multi-target tracking, the unification of multi-target tracking in sparsity and multi-target tracking in high density, and the unification of target tracking and situation perception; (2) it establishes the group-target tracking architecture; (3) it proposes the method to realize group splitting detection and group-target track initiation simultaneously; (4) it proposes for the first time the nearest neighboring—all-neighboring single group-target association algorithm; (5) it proposes for the first time the multi-group-target association algorithm under double multi-multicorrespondence; (6) and it proposes for the first time the method to realize group-target combination/separation detection, situation synthesis, and intention judgment simultaneously. To highlight the themes with limited words, the book does not discuss in detail the so-called fundamental theory on target tracking and instead, only provides some relevant basic knowledge. The readers are referred to other references when more information is needed.

My research on group-target tracking is a long process. Based on my actual work experience, I came up with an idea, asked experts for advice and guidance, and then tried many methods to work on it before the final complete algorithm system is shaped. During this period, I drew inspirations from the works of Hongren Zhou, an expert in information technology and executive chairman of Advisory Committee for State Informatization, received helpful guidance from academician De Ben, researcher Qingyu Cai and professor Yingning Peng, and benefited from the valuable suggestions of Ms. Hongya Liu. I would also like to express my heartfelt thanks to Mr. Jun Lu and Jianliang Zhu for their efforts to check on relevant issues.

However, target tracking theories and technologies change with each passing day. While I have referred to massive materials and used a lot of research achievements for reference, it is impossible to avoid defects and mistakes in the book due to insufficient knowledge reserve and incapability. Therefore, I'm looking forward to the generous opinions and advice of experts and scholars that are focused on target tracking.

The book is translated from Chinese by Ge Geng and Fan Yang.

Thanks to National Defense Sci-tech Publishing Fund for its support.

Thanks to Mr. Zhili Xiao for his help and guidance to the publishing of the book.

Thanks to all experts and scholars that have contributed to the book directly and indirectly.

Thanks to my family for their constant support over the years.

Beijing, China   Wendong Geng
June 2015

Contents

Abstract

It is always an attractive subject to realize the unification of single-target tracking (STT) and multi-target tracking (MTT), the unification of sparse multi-target tracking and high-density multi-target tracking, and the unification of target tracking and situation perception. With this in mind, the authors come up with the idea of group-target tracking.

This book elaborates the general situation of target tracking technologies at both home and abroad, as well as the research results of the authors in this field. It covers seven chapters, with main contents including general development of target tracking and typical tracking systems; background for group-targets, formation process of group-target tracking theory, group-target concept and connotation, group-target tracking principles, and their application prospects; basic concepts of target tracking, Kalman filtering, target motion model, target tracking algorithms, and other basic knowledge; group splitting detection for the formation of group-targets and group-target track initiation algorithms; single-/multi-group-target association algorithm based on integrated data association and situation association; and group-target combination/separation detection algorithm enabled simultaneously by maneuver detection and situation perception, as well as simulation design and verification of group-target tracking algorithms, among others.

This book can be used as a reference by technical personnel engaged in such areas as radar data processing, MSDF, electronic countermeasure, situation perception, and command information system, as well as textbook for undergraduate and graduate students studying relevant majors.

Chapter 1
Introduction

1.1 Foreword

Target tracking is an essential issue in the real world both for the survival of animals and for the life of human beings. For instance, to capture a prey, a lion must fix its eyes all long on the target. In military and civil fields such as ballistic missile defense, air defense early-warning, sea and ground battlefield surveillance and air traffic control, tracking targets reliably and precisely is pursued all the time by sensor tracking systems like radar.

With the development of modern detection technology, advancement of weaponry, and increasing complexity of targets and environments, target tracking system is faced with many new problems and new challenges.

With regard to targets, there are multiple challenges, such as diversity of antidetection means, complexity of maneuvering manners, and density of attacking multi-targets. With regard to detection environment, increasing interference from the nature and growing complexity of man-made interference are some of the challenges. With regard to detection equipment, multi-returns exist both in multi-targets and in extended single targets. With regard to battlefield situation, large quantity, multi-functionality, and interleaving of battling targets are some of the problems to address. Therefore, modern target tracking systems face not only estimation of motion state of extended targets and high-density multi-targets amid interferences, but also target tracking under complicated battlefield situation as a result of various forces intervening together, including enemy, friend, and our own force. Additionally, for each force, the situation description of the other forces is still a problem.

This chapter first introduces the basic situation of target tracking, typical multi-target tracking system, and basic problems faced. Then, it describes the background for emergence of a group-target, puts forth group-target concept, specifies group-target tracking idea, and elaborates the fundamental principles of group-target tracking.

© National Defense Industry Press and Springer Science+Business Media Singapore 2017
W. Geng et al., *Group-target Tracking*, DOI 10.1007/978-981-10-1888-6_1

1.2   Target Tracking Overview

Target tracking technology can be classified by type of sensors, by tracking method, or by quantity of targets under tracking. By quantity of targets, target tracking can be divided into single-target tracking (STT) and multi-target tracking (MTT). STT includes single-target tracking by single sensor and single-target tracking by multi-sensors. MTT includes multi-target tracking by single sensor and multi-target tracking by multi-sensors. Multi-sensor tracking system is also called data fusion tracking system, or information fusion tracking system.

1.2.1   STT Profile

In 1795, famous mathematician Johann Carl Friedrich Gauss used least squares method for the first time to estimate the track of the Ceres. This pioneered a realm of science that applies mathematical method to process observation and experimental data. Research on STT can date back to the eve of World War II, when SCR-28, the first tracking radar in the globe, came into being in 1937, which started the radar development history. As a one-to-one closed-loop tracking radar, SCR-28 adopts gate range tracking and cone scanning angle tracking. With the optical axis of the sensor always pointing at the target it tracks, SCR-28 implements tracking based on closed-loop system's error elimination principle. Also in that period, the Wiener filtering theory raised by Norbert Wiener (1894–1964) et al. initiated a new research field—modern filtering theory (based on probability theory and theory of random processes). As soon as it came out, Wiener filtering theory was immediately applied in telecommunications, radar, and many other areas with successful results. Due to low precision and weak anti-interference capability, however, this type of radar could not meet the increasing needs of wars. Therefore, monopulse radar, a milestone in the development of STT radar, emerged in 1950s. With the development of digital technology and estimation theory after 1960s, digital tracking systems emerged. In terms of theory, Kalman filtering theory and nonlinear filtering method went through unprecedented development. In early 1970s, Singer et al. came up with a series of maneuvering target tracking methods, and then in the middle of 1970s, Pearson and Shibata et al. succeeded in applying Kalman filtering technology in airborne radar tracking system. These achievements were making STT technology increasingly mature. Along with weaponry development and increasing war complexity, tracking of single target by radar is no longer able to meet the needs of wars. As a result, all countries are launching research and development of technologies for tracking multi-targets by single sensor [19] while optimizing STT.

STT mainly involves measurement data preprocessing, maneuver model selection, maneuver detection and maneuver identification, filtering and prediction, and selection of tracking coordinate systems and filter state variables, among other elements, as shown in Fig. 1.1.

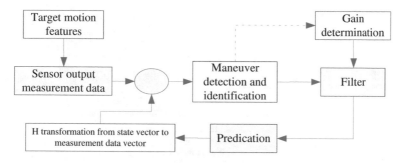

Fig. 1.1 STT architecture

1.2.2   MTT Profile

The basic concept of MTT was put forward by Wax in his article published in 1955 on the Journal of Applied Physics. He considered MTT an issue similar to determining a particle path amid noises in atomic physics. Based on this, Wax worked out the basic procedures of target tracking that covers track initiation, track maintenance, and track termination. In 1964, Sittler published a thesis on IEEE entitled Optimal Data Interconnection in Surveillance Theory, which became the pioneer work of multi-target tracking. In his paper, Sittler presents the Bayes method featured by the association of multi-target observation data with optimal track data, laying a foundation for later research (he used track furcation method at that time as Kalman filtering had not been widely applied yet). In early 1970s, Bar-Shalom and Singer introduced Kalman theory and other relevant theories into the MTT field, resulting in flourishing development of MTT research with a mass of articles and reports published. The nearest-neighboring data approach (NNDA) put forth by Singer is the simplest method to address data interconnection (but this method has relatively low accuracy in a common noise wave environment). Bar-Shalon came up with probabilistic data association (PDA) algorithm for STT in 1975 [28], which is highly suitable for noise wave environment. Thereafter, Bar-Shalom and his students put forward joint probabilistic data association (JPDA) algorithm [24, 30], on the basis of PDA that is only fit for STT. IPDA is currently one of the ideal approaches to address intersection of tracking gates of multi-targets being tracked, and obtaining the joint association probability of every measurement data is the core concept of this theory. The golden age for the development of MTT was in 1970–1985, during which considerable researches were conducted on MTT technologies centered on radar in foreign countries, with abundant achievements made in maneuvering target tracking, data association methods, and other aspects. In the middle and late period of the twentieth century, the focus of research in foreign countries shifted from multi-target tracking by single sensor to multi-target tracking by multi-sensors. Later on, research on radar target tracking started shifting from single radar to multi-radar, from multi-radar to multi-sensor tracking. As a result, starting from late 1970s, an emerging subject—multi-sensor data fusion

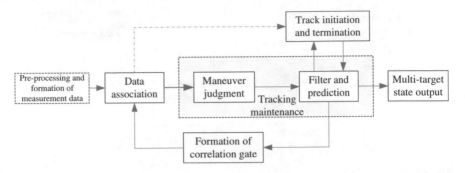

Fig. 1.2 MTT architecture

(MSDF) grew quickly and has so far been applied extensively in modern C^3I system, various weaponry platforms, and many civil fields. In a word, the development history of target tracking generally has covered single-target tracking by single radar, multi-target tracking by single radar, single-target tracking by multi-radars, multi-target tracking by multi-radars, and information fusion with dissimilar multi-sensors.

It is worth noting that the maneuvering performance of targets like aircrafts keeps improving and target tracking environment is deteriorating. As a result, much higher requirements are raised for the tracking capability of sensors like radar, whenever they are either used alone or used as a basic unit of MSDF. Even if sensors like radar are used as a basic unit of MSDF, poor tracking quality will severely impact information fusion quality and increase unnecessary calculation. Therefore, target tracking quality of sensors like radar has become extremely important.

In addition to the essential elements of STT, MTT also needs many other elements, such as formation of tracking gate, data association and tracking maintenance, track initiation and termination, and processing of false alarm and missing detection. And its tracking principle is as shown in Fig. 1.2.

1.2.3 Data Fusion Profile

With the development of science and technology, the performance of sensors has seen significant enhancement and diversified multi-sensor systems for complicated application background have emerged in large numbers. Since 1970s, in particular, the emergence of precision-guided weapons and long-distance attack weapons has extended the battlefield to 5D space covering land, sea, air, aerospace, and electromagnetism. To achieve the best effect of military operations, multi-sensors must be used to provide a variety of observation data, as the approach of relying only on single sensor to provide data can no longer meet the needs of military operations in the new-generation combat system. These sensors include various active and passive probes that cover broad frequency bands, such as microwave, millimeter wave (MMW), TV, infrared, laser, electronic support measures (ESM), and electronic intelligence (ELINT). After optimized integration and processing of these data, multi-sensor integration enables to obtain useful combat information in a real-time manner, including target discovery, state estimation, target attribute, behavior intention, situation assessment, threat analysis, fire control, precision guidance, electronic countermeasure, combat mode, and assistant decision-making. In multi-sensor system, as the form of data is diversified, the amount of data is huge, data relationship is very complicated, and data need processing in a real-time manner, it goes beyond the capacity of human brain to comprehensively process such massive and complex information. As a result, MSDF quickly gained traction and has so far been applied extensively in modern C3I system as well as various weaponry platforms and many civil fields.

Over the past 20 years, MSDF technology has been increasingly attracting public attention, with the word "fusion" being very frequently cited in both military and non-military fields [1-0]. Military application involves sea surveillance, air-to-air and ground-to-air defense, battlefield monitoring, target detection, and strategic early warning, among others. Non-military application covers law enforcement, remote sensing, automatic monitoring of equipment, medical diagnosis, robot technology, and so on. Information fusion is a new approach to process information based on the specific issue that uses multiple and/or multi-type sensors in one system, which is also called multi-source association, or multi-source synthesis, or sensor blending, or multi-sensor fusion, but more widely known as multi-sensor information fusion or data fusion. At present, it is difficult to have a unified definition for information fusion, due to the extensiveness and diversity of the contents studied. Since the Gulf War, especially Kosovo War, there are a growing number of people engaged in the research of information fusion technology with a lot of new achievements. In some sense, information fusion can be roughly summarized as a data processing process that uses computer technology to automatically analyze, optimize, and integrate the observation data of several sensors obtained according to timing sequence to accomplish the required decision-making and estimated task. In accordance with this definition, various sensors are the foundation of information fusion, multi-source information is the processing target of information fusion, and coordinated optimization and comprehensive processing is the core of information fusion [1-9]

In the military sector, information fusion mainly includes detection, interconnection, association (relevance), state estimation, target identification, situation description, threat estimation, sensor management, and database. It is a process in which sensor data are processed comprehensively at multiple levels. Each processing level mirrors the abstract of original data to varying degrees. Information fusion is a complete process from detection to threat judgment, to weapon distribution and to channel organization. The result is that the state and attribute are assessed at the lower level and the whole situation is assessed at the higher level. This, in some sense, emphasizes that the core of information fusion is to process the data from multiple sensors at multiple levels, multiple aspects, and multiple layers, so as to generate new, meaningful information that cannot be obtained by any single sensor [9–11].

To sum up, the so-called information fusion means comprehensive processing of information from multiple sensors or multiple sources to obtain more accurate and reliable conclusion. Another common name for information fusion is data fusion. As far as the connotation of information and data is concerned, however, the term "information fusion" is more popular, accurate, reasonable, and generalized. It is widely believed that information contains not only data, but also signal and knowledge. That is why the term "information fusion" is widely used internationally.

In fact, MSDF is a basic function of humans and other creatures. Humans can instinctively integrate the information (scenery, sound, smell, and touch) detected by various functional organs (eye, ear, nose, and limbs) with transcendental knowledge, so as to estimate the surrounding environment and what is happening. MSDF is a simulation of the function of human brain to comprehensively process complicated issues. Like human brain, in principle, it makes the full use of multi-sensor resources. By rationally allocating and using various sensors and the information they have observed, it combines the complementary and redundant information of various sensors in time and space based on certain optimization criteria so as to generate consistent interpretation and description of the environment observed. The ultimate aim is to improve the effectiveness and efficiency of the whole sensor system through joint operation of multiple sensors.

Moreover, single-sensor signal processing or multi-sensor data processing at the bottom layer is a simulation of human brain information processing at the bottom layer. On the contrary, MSDF is to obtain to the largest extent the amount of information about the targets and environment being detected by effectively using multi-sensor resources. MSDF also differs essentially from the classic signal processing method in that the multi-sensor information processed by the information fusion system has more complex form and often appears in different information layers. These information abstraction layers include signal layer, data layer, and perception layer.

MSDF has multiple advantages in addressing detection, tracking, target identification, and other issues. To be specific, it boosts system survivability, expands space and time coverage, improves credibility, reduces vagueness of information, optimizes detection performance, enhances spatial resolution, reinforces system reliability, and increases dimensions of the space measured. Compared with single-sensor systems, multi-sensor systems have greater complexity, which causes some negative factors, such as increases in cost, in system size/weight/power consumption, and in the probability for the system to be detected due to more radiation. As a result, it is necessary to comprehensively balance the performance gains and negative factors brought by multi-sensors.

In a word, MSDF has grown into a hot research area and is currently a key common technology concerned by many countries. With technological advancement, the MSDF technology is set to play greater role in military, civil, and other sectors.

1.2.4 Formation Target Tracking Profile

(1) General situation of formation target tracking in foreign countries

G. Binias presented the concept of formation tracking in his article entitled Report of the FFM [10] in 1977 and published his thesis entitled Computer Controlled Tracking in Dense Target Environment Using a Phased Array Antenna [13] at IEEE Conference in the same year. He called the system "computer-controlled target tracking" on the basis of the multi-functional array radar system experiment by Wachtberg-Werthhoven, West Germany. The working process of the system is as follows. (a) Use a regular, spatial orientation search program to detect unknown targets. (b) Give tracking commands based on the acquired search map (space-time coordinates of target return) to guide the antenna beam to the search map for further search for the target return. (c) Continue tracking according to the tracking command and guide the antenna beam to the predetermined target location. In March 1978, G. Binias submitted a report entitled Basic Theory and Formation Track Initiation [11], presenting the formation tracking concept and discussing formation track initiation, among other issues. In June 1979, he submitted another report entitled The Formation Tracking Procedure for Tracking in Dense Target Environment [12]. The report discusses tracking of formation center and formation margin, studies formation track initiation, formation track maintenance, and processing of formation combination and separation, and proves the effectiveness of formation tracking through actual formation flying test.

Erwin Taenzer submitted a report entitled Tracking Multiple Targets Simultaneously with A Phased Array Radar [8] in 1977 and published a same-titled thesis [9], introducing approaches to track high-density multi-targets by radar. According to his article, the targets being tracked can appear in the form of formation, maneuvering or overlapping with each other, can fade away or reappear, and can or cannot be identified. He came up with typical requirements for formation tracking, discussed fundamental principles, and provided test results.

E.H. Flad published an article entitled Tracking of Formation Flying Aircraft [14] at International Radar Conference 1977 held in London. In his article, he analyzes the mechanism of fuzzy data association in MTT and describes the information that formation tracking can provide (e.g., the location and structure of targets in formation, as well as the number of targets contained in the formation and the speed of targets) and association methods for formation tracking, among others. In 1985, A. Farina and F.A. Styder from Italy published the book Radar Data Processing [16], which involves the concept of formation target in Chapter 4 and offers a function flowchart as shown in Fig. 1.3. Shyu, H.C., Lin, et al. published an article entitled The Group Tracking of Targets on Sea Surface by 2-D Search Radar [15] at International Radar Conference 1995, proposing for the first time the cluster-seeding group initiation concept. What is worth noting is that the book Pattern Recognition Principles written by J.T. Tou and R.C. Gonzalezez in 1974 [17] provides the K-means to get the target cluster center, and the article The Group Tracking of Targets on Sea Surface by 2-D Search Radar [15] also considers this

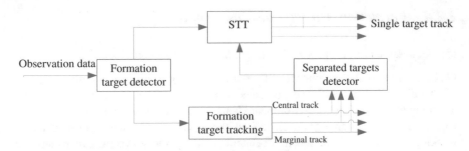

Fig. 1.3 Formation target tracking architecture

method a typical approach to obtain the cluster center. In addition, A.P. Frazier and J.A. Scott came up with the algorithm for cluster tracking of moving targets in their article An Algorithm for Tracking of Moving Sets published in 1976, whose basic idea is tracking the center of a moving cluster with the same speed and direction.

According to the above analysis, formation target tracking involves the following main contents: group splitting, formation target track initiation, formation target data association and tracking maintenance, combination/separation detection of formation target and its members, etc.

(i) Group splitting

Group splitting of formation target can be summarized in two categories—observation data-based group splitting and track-based group splitting.

Observation data-based group splitting is implemented by computing the distance between observation data one by one. First set threshold value T, and compare the distance between each of the two measurement data; then, the set of measurement data with distance less than T constitute a formation target. When multiple formation targets (known as subgroups) exist in a surveillance area, the group splitting processing method first determines the distance d_{ij} between all members of any two subgroups and lets $d_0 = \min [d_{ij}]$; when $d_0 > T$, the two subgroups become independent formation targets.

Track-based group splitting is a method to process group-targets after every target track is obtained. The first step is to set track space T and target density threshold L; the next step is to segment surveillance space into several units based on $\frac{T}{3} \times \frac{T}{3}$ and calculate the number of tracks of each unit; the multi-targets with track space smaller than T and density larger than L constitute a formation.

(ii) Track initiation

Similarity method and furcation method, K-means, and cluster-seeding track initiation method are proposed in such works as *Phased Array Radar Data Processing and Simulation Technology* and Pattern Recognition Principles [2, 17], which will be explained below.

Similarity method and furcation method for track initiation introduced in *Phased Array Radar Data Processing and Simulation Technology* [2] are based on the formation target. According to these methods, the speed and direction of members in a subgroup are fundamentally identical, and the same filter is used, meaning the filter gains of all tracks in a subgroup are the same. The requirement for almost same speed and direction of members in a subgroup is limited to track initialization stage, and there is no such requirement after the tracking becomes stable.

Similarity method is used when there are three or more targets. Take any two measurement data in the K cycle and draw a line between them, find the corresponding measurement data lines that are basically parallel and roughly equal in length in the $K + 1$ cycle, and then estimate the speed of formation target based on two pairs of measurement data obtained in the adjacent cycles. Furcation method is used when there are two targets in the formation, with almost same principles as the similarity method. The difference is that three scanning cycles are needed to complete a track initiation when there are multiple subgroups (no more discussion here).

J.T. Tou and R.C. Gonzalezez introduce K-means and cluster-seeding method for initiation in their work *Pattern Recognition Principles* [17]. K-means takes the following steps. Take the first sample as cluster center; obtain the return that has the smallest distance from the cluster center among all sampling values, and merge it into K cluster; obtain the new center that includes the return merged; and repeat the above process with the new center to enable all returns to be combined into K cluster successively until the central value converges on a certain value. Cluster-seeding method takes the following steps. Take the first sample as cluster center; set a threshold value T; calculate the distance between the sampling value of next cycle and all cluster centers of the current cycle, and if a sampling value is larger than T, it will be used as new cluster center; otherwise, merge into the cluster with the smallest distance, and the clusters that merge returns establish new cluster centers. The efficacy of cluster-seeding method depends on threshold value [17], and the value usually depends on the features of a cluster or requirements of a mission. The advantages of cluster-seeding method include high accuracy, short computing time, and small interference effect as well as no need to know the number of clusters.

It is worth noting that the group splitting processing and similarity track initiation introduced in the book *Phased Array Radar Data Processing and Simulation Technology* by Qingyu Cai, Yi Xue and Boyan Zhang [2] take asynchronous actions, that is, track initiation after group splitting processing. K-means and cluster-seeding method combine group splitting processing and group initiation by taking a synchronous action, which will have important reference significance to group splitting detection and group initiation.

(iii) Data association and tracking maintenance of formation targets

Publications by E. Taenzer, Erwin Taenzer and G. Binias [8–13] describe the formation-based formation target tracking method from different perspectives. In nature, this method performs center-based tracking and margin-based tracking. Formation tracking needs three tracks—central track of a formation target and two marginal tracks. When formation tracking kicks off, determine which single targets can constitute a formation target through agreed target distance, and separate them from the environment they are in. After getting the formation target, the next step is to calculate the average location of all targets, then control a beam to irradiate that location so as to obtain the measurement data required by the central track, and set up the central track of the formation target. Meanwhile, find the location of the marginal targets in the coordinates, and set up marginal tracks based on the measurement data of the two marginal targets. Formation tracking is thus enabled based on these three tracks.

As for the data association of the three tracks of the formation target, the book *Phased Array Radar Data Processing and Simulation Technology* [2] applies the nearest-neighboring association algorithm, while *A New Algorithm for Group Tracking* [6] adopts multi-hypothesis algorithm. In other words, data association for formation target tracking applies the association algorithm of traditional multi-target tracking.

(iv) Combination and separation of formation target members

Combination and separation of formation target members are judged by tracking the quantity and location of targets of the marginal targets within the wave beam. When several targets appear within the marginal beam, it is necessary to timely remove the targets closer to the formation center and retain the targets farther to the formation center, and meanwhile, the size of the formation target becomes larger. Conversely, members of the target are separated from formation after they leave the formation margins.

A method to combine and separate old and new members in a formation target based on observation data and track is introduced in *Phased Array Radar Data Processing and Simulation Technology* [2]. This method associates measurement data obtained in every radar cycle with formation members one by one. Then, the measurement data that are associated successfully under certain threshold will maintain, and those not associated successfully will be handled separately. Supposing the radar receives M effective returns in one scanning, then measurement data $Y_i(k) = (R_i, \alpha_i, \beta_i)$ $i = 1, 2, \ldots, M$.

Supposing there are N members originally in the formation, then associate M measurement data $Y_i(k)$ with N members one by one for association judgment. The formula is as follows: $\delta(i,j) = Y_i(k) - H\widehat{X}_j(k/k-1)$ $(j = 1, 2, \ldots N;$ H for measurement data matrix). If $\delta(i,j) \leq X_G$, then return i is correlated with member j. In the formula, $\widehat{X}_j$ $(k/k-1)$ refers to the predicted value for member j in the formation. X_G is the threshold determined by measurement error and prediction error. If multiple returns fall in X_G, the nearest-neighboring method will be used. The process continues until all the M measurement data are associated with N formation members.

If measurement data correspond to formation target members successfully, then separation check starts. Distance separation threshold is set to be R_0 (depending on the design parameters of signal processor and false-alarm rate), then separation is confirmed when $|R_c(k) - R_i(k+1)| > R_0$, where $R_c(k)$ is the distance measurement data of formation target centroid and $R_i(k+1)$ is the distance measurement data of return i. As for judgment of angle separation, first set an angle separation threshold Δ_0, then figure out the angle value of central track of the formation, and then subtract the angle value of every measurement data. As a result, those targets with greater angle difference value than Δ_0 are judged to have angle separation.

Formation target tracking mainly includes group splitting, track initiation, data association, and separation processing, with its tracking principles shown in Fig. 1.3.

(2) General situation of formation target tracking in China

In 1991, Hongren Zhou, Zhongliang Jing, and Depei Wang published the book *Tracking of Maneuvering Targets* [1]. The work mentions that the algorithm for obtaining equivalent statistical center via probabilistic data association can be applied in target group tracking, but does not give specific algorithm. In 1988, Qingyu Cai and Liancheng Wang published the work *Data Processing Method for Tracking of Flying Targets in Formation by Phased Array Radar* [3], and in 1991, Peizhang Jia and Liancheng Wang published the work *Method for Tracking of Multi-targets in High Density* [4]. Relevant contents in the two books are reflected in the book *Phased Array Radar Data Processing and Simulation Technology* published in 1997 by Qingyu Cai, Yi Xue, and Boyan Zhang. These works provide the methods for group splitting based on observation data, track initiation, and formation target tracking, and specify the rules for combination and separation of formation target members. In 1995, Chuwei Sun and Tanlin Yuan compiled the book *Group Tracking of Sea Targets by 2D Search Radar* [5]. In 2001, Huaili Wang et al. delivered a paper entitled *A New Algorithm for Group Tracking* [6] at Beijing International Radar Conference 2001. It mainly introduced ways to reduce computation burden by means of track-based group splitting on the premise of multi-hypothesis algorithm and tells how formation target is formed with given target distance and density. In 2006, Wendong Geng, Hongya Liu, et al. delivered an article entitled Study of Kalman-Based Algorithm for the Maneuvering Group-Target Tracking at Shanghai International Radar Conference 2006. For the first time, it puts forth the concept of group-target in the sense of data association, comes up with a new algorithm that calculates the weight of every return in each tracking gate by using the ratio of Kalman filter residual vector norm to tracking gate threshold, provides a formula to obtain equivalent measurement data, and establishes the Kalman equation for group-target tracking.

1.2.5 Typical Target Tracking System

(1) Monopulse radar

Monopulse radar is precise tracking radar with automatic angle measurement system. Like other pulse radars with automatic angle measurement system, it emits a series of pulses instead of one pulse. However, monopulse radar uses simultaneous lobe method to measure angles. In theory, full-angle information can be obtained just by comparing the same return pulse that each wave beam receives. This is how the name "monopulse" comes. By the way how the angle error signal is extracted, monopulse radar is generally classified into amplitude and sum–difference monopulse radar as well as phase and sum–difference monopulse radar. The principle for angle measurement is as follows. Within the same time span, multiple antenna lobes deployed on both sides of the equisignal axle receive returns simultaneously and compare them to obtain angle error signal; then, after being amplified, the error signal drives the antenna to move toward the direction with reduced error.

(2) Bistatic/Multi-static radar

Bistatic radar (BR) refers to such radar that has separate transmitting antenna and receiving antenna and is also the earliest basic pattern in the world. Monopulse radar did not appear until the invention of transmitting/receiving (T/R) diode. If a transmitting station has multiple separate receiving stations, or multiple transmitting stations coordinate with multiple receiving stations, such system is called multi-static radar. If the transmitting station is also able to receive return signals of the target, such system is also called composite multi-static radar. As receiving station is passive, it has good capability to resist the attack of antiradiation missiles and is also important antistealth radar. A typical multi-static radar system, for instance, is AN/FPS-133 surveillance fence radar that has three transmitting stations and six receiving stations.

(3) Tracking while scanning

Tracking while scanning (TWS) is such radar that can simultaneously conduct mechanical scanning at constant rotation speed, wave beam searching, and target tracking. Its tracking system is generally composed of track initiation, observation data–track association, tracking filter, track management, etc. This kind of radar usually divides a search space into several sections, and observation data–track association is completed within one section and within two adjacent sections, instead of association in the entire space. The size of a section depends on such factors as scanning cycle, speed of targets, density of targets, and observation errors.

(4) Phase array radar

As an epoch-making radar, phase array radar (PAR) has many advantages over radars using mechanical scanning. First, it features multi-target searching and tracking as well as multiple radar functions, including tracking and searching (TAS), TWS, searching by section, and burnt out. Second, it is able to change searching data rate and tracking data rate. Third, it has the ability of adaptive spatial filtering and adaptive space-time processing. Fourth, it enables high-power aperture product and variable aperture product. Fifth, it enables conformal antenna aperture and radar platform.

(5) Multi-sensor tracking system

One of the key factors to multi-sensor tracking system is the data fusion technology. By number of targets being tracked, it includes multi-sensor single-target tracking system and multi-sensor multi-target tracking system. By type of sensors, it includes same-type multi-sensor target tracking system and different-type multi-sensor target tracking system. By information processing method, it includes centralized tracking system and distributed tracking system. By layer of information fusion, it includes signal layer fusion tracking system, data layer fusion tracking system, and situation layer tracking system. Common multi-sensor multi-target tracking systems include radar network, ballistic missile defense surveillance network, space target surveillance network, shipborne air defense radar system and space station surveillance system. Many new names are appearing now, such as network radar and network-centric warfare-based track combination, but none of them separates from the target tracking issue based on multi-sensor information fusion. Multi-sensor tracking systems have many advantages, such as strong   anti-interference   and   antidestroying   ability,   wide   space-time frequency-domain coverage, excellent robustness, and high measurement accuracy.

1.3   Group-Target Tracking Theory

1.3.1   Background for the Emergence of Group-Target Tracking

(1)   Traditional multi-target tracking algorithms no longer meet the requirement of tracking multi-targets in high density

In the multi-target tracking field, typical methods mainly include the nearest-neighboring data association (NNDA), probability data association (PDA), and joint probability data association (JPDA). NNDA [35, 37] enables multi-target tracking by determining the correct measurement data based on the principle of smallest distance between measurement data and data from the predicted center. It is only applicable to tracking of multi-targets in sparsity and cannot meet the requirement of tracking multi-targets in high density. PDA [20–23] applies to single-target tracking amid high-density returns, with an assumption that the correct measurement data in each tracking gate are unique and others are regarded as false measurement data obeying uniform distribution. If the measurement data of another target fall into its tracking gate with lower probability (i.e., when the distance between targets is not so small), it is tolerable to handle the interference as uniformly distributed false measurement data. However, once the distance between two targets is so small, each target causes continuous interference to other targets when the track of a target enters the tracking gate of the other. Such interference often results in unstable tracking or mistaken tracking. Therefore, this method is also not competent enough for tracking of multi-targets in high density.

JPDA, built on PDA, is widely recognized as a classic algorithm for tracking multi-targets amid high-density multiple returns. It tracks multi-targets through joint probability data association by considering all targets in the surveillance area.

However, there are two preconditions. The first one is that the number of targets is known, that is, target tracks have been initiated. The second one is that every measurement data can only be sourced from one single target, and each target can have one measurement data at most. That is, each measurement data correspond to each target. As we know, one of the problems in JPDA is overload caused by excess computing due to increased number of targets [1]. To solve this problem, researchers provided multiple simplified JPDA algorithms and conducted studies on parallel computing to deal with high-burden computation. Despite some advantages, these simplified algorithms will surely lead to loss of some performance, and large-scale parallel computing needs more software and hardware resources and costs. In the case of targets in high density, it is hard to guarantee one-to-one correspondence of each measurement data with each target when the locations between targets are close to a certain extent. This lack of one-to-one correspondence leads to the failure of JPDA. That is to say, JPDA as well as its enhanced algorithms still cannot meet the requirement of multi-target tracking in high density, despite its contribution to the realization of multi-target tracking amid high-density returns.

(2) The formation tracking algorithm for formation targets does not have universal meaning

As defined in relevant publications [2–8], the purpose of formation target tracking is to address tracking of formation targets in the context of high-density targets and to save limited radar and computer resources, especially when there are more than three targets. The principle of formation tracking is on the basis of mean cinematical behavior. Depending on center-based tracking and margin-based tracking, center-based tracking realizes tracking of the formation as a whole, while margin-based tracking realizes tracking and separation processing of marginal members of the formation, so as to timely adjust the scale of the formation to ensure stable tracking of formation targets. That is to say, formation tracking is about correct tracking of formation targets with strict location and movement restraint. The background under which it is applied and its composition form determine that the tracking method is only applicable to the special case of formation target, among cases of multi-targets in high-density, and does not have universal meaning for tracking of targets in high density.

(3) Analysis on the forming process of typical high-density targets

Take a foreign plane model as an example. Two planes fly abreast, and each launches a missile at the same time. The plane opens the missile cabin to let the missile have free fall, engine ignites in a few seconds, and then the missile starts accelerated movement and leaves the planes. In the process, after the missiles are launched, they are combined with the planes to form a high-density multi-target group. When the missiles leave in acceleration after ignition, this high-density multi-target group is split to form two low-density multi-target group composed, respectively, of missiles and planes. After flying for a period of time, the two planes and the two missiles form four single targets. As the missiles fly at almost same speed with the planes in horizontal direction and have free fall in vertical direction after they are launched and before their engines ignite, their distance from the planes is getting larger and larger from zero. During this period, therefore, the missiles and the planes jointly form a high-density multi-target group, making it hard for radars to distinguish the returns of the planes from that of the missiles. Moreover, once the engine ignites, the missiles will leave the target group with an extremely large acceleration and fly a very short time. It is these factors, including returns which are hard to distinguish and high acceleration after ignition of missile engine, that make it hard for radars to capture the missiles in time. Subsequently, the ballistic trajectory of the missiles cannot be obtained, which will result in extremely severe consequences in war. This once again demonstrates the importance and necessity of researching on tracking of multi-targets in high density. After all, once the challenges involving high-density multi-target tracking featured by separation of plane from missile are tackled, problems related to high-density multi-target tracking involving plane formation will be readily solved.

The above analysis shows that the multi-targets resulted from separation of plane from missile are multi-targets in high density but not formation targets. As a result, traditional multi-target tracking algorithms and formation target tracking algorithms are unable to solve the practical problems. Therefore, it is necessary to free our minds to find new methods to solve the problems related to high-density multi-target tracking.

1.3.2   Formation of the Group-Target Tracking Theory

(1) Classification of multi-target tracking

In early studies, multi-target tracking (MTT) refers to tracking of single targets in an environment with multiple noise waves [1]. In a real sense, however, MTT considers not only returns from multiple targets, but returns from targets that are not being tracked, noise waves, false alarms, fixed targets, and other interferences. MTT is aimed to break up the measurement data received by radar into measurement data sets corresponding to different targets, namely tracks. Once the track of each target is formed and confirmed, the number of targets tracked by radar and motion parameters of each target, such as location, speed, acceleration, and even characteristics of target, will be estimated. In *Tracking of Maneuvering Targets* [1], target tracking amid multiple returns is classified into four categories—tracking of one target by one sensor (OTO), tracking of multi-targets by one sensor (OTM), tracking of one target by multi-sensors (MTO), and tracking of multi-targets by multi-sensors (MTM).

Based on the above classification, the book further classifies tracking of multi-targets into the following five scenarios.

First scenario: Tracking of sparse multi-targets. When sensors can identify and associate targets without vagueness, we usually adopt the nearest-neighboring data association (NNDA).

Second scenario: Single-target tracking (STT) under high-density multi-returns. We usually adopt all-neighboring probability data association (PDA).

Third scenario: Multi-target tracking (MTT) under high-density multi-returns. We commonly adopt joint probability data association (JPDA) [22, 25, 28]. This algorithm was put forward by Bar-Shalon and his students based on PDA that is only applicable to STT. It is relatively a more effective way to solve tracking of multi-targets when association gates intersect, although it needs further improvement.

Fourth scenario: Tracking of multi-targets in high density. Some publications mix this case with MTT amid high-density multi-returns, and in fact, they are two different concepts and different scenarios. High-density MTT refers to such a multi-target scenario that targets are hard or impossible to distinguish due to small target distances. A typical example is aircraft formation targets shaped during flying. It features not only numerous returns and small distance between formation members, but also serious intersection of association gates making it extremely difficult to enable correct association.

Fifth scenario: Tracking of multi-targets in high density amid high-density multi-returns. For example, high-density multi-returns are generated when aircraft formation fly in low altitude, and when shipborne weapons launch multiple bombs/missiles simultaneously and then bombs/missiles separate from the ship.

As a matter of fact, measurement of the sparsity and density of multi-targets is closely related to the resolving power of radar, rather than absolute quantity of the spatial distance between targets. Moreover, the density of targets is also related to the relative location between target and radar, and changes with time and space. Therefore, reliable and stable tracking of targets also depends on radar's resolving power in addition to its energy and time resources [1, 2]. The resolution of early radars for tracking of single targets relies mainly on the time width and beam width of radar signals, and radars are unable to distinguish multiple targets within their beam scope. The resolution of MTT radar using data association technology depends mainly on the size of tracking gate, but the tracking gate always occupies certain space. As a result, when it is smaller than the tracking gate, the target distance makes it difficult or impossible for radar to reliably and stably track every target, which is a great challenge radar faces in high-density multi-target tracking.

(2) Formation of the group-target tracking idea

As mentioned above, multi-targets generally include sparse multi-targets (each target is called identifiable single target), multi-targets amid high-density multi-returns, and high-density multi-targets. At present, single radar can track multi-targets in sparsity, and the algorithm for multi-target tracking amid high-density multi-returns is getting increasingly mature. However, single radar cannot effectively track multi-targets in high density. High-density targets usually refer to multi-target assembly that features very small relative spatial distance between targets in a certain period, and the multi-target assembly is hard or impossible to be identified by radar. The reason for difficult tracking is the small target distance which causes vague data association. Currently, all-neighboring data association algorithm represented by JPDA and formation tracking algorithm for formation target tracking are likely the most applicable algorithms to tracking of multi-targets in high density. As multi-target tracking algorithms like JPDA have such problems as mistaken or missing tracking when targets are in high density, scholars are trying their best to improve data association algorithm to enhance the capability to track multi-targets in high density. However, the results of these efforts are limited, as the tracking gate of radar cannot be infinitely small. Formation target refers to multi-target assembly of which targets have very small target distance but basically same speed and direction. Formation target is a special form of high-density multi-target, so formation tracking algorithm only solves a particular case in high-density multi-target tracking and does not have universal meaning. In other words, these algorithms and their improvements have not solved the problems of tracking multi-targets in high density.

We can draw the following conclusions based on further analysis on current data association technologies and formation target tracking methods. (i) The technology for tracking sparse multi-targets is mature. (ii) Multi-target tracking algorithms amid high-density multi-returns have achieved fruitful results. (iii) The resolving power of traditional data association has relative physical limit and cannot reliably and stably track every target within such limit. (iv) Formation target tracking only solves a particular case in high-density multi-target tracking, without universality. In another word, there is currently no effective method to track multi-targets in high density.

Now that technologies for tracking of sparse multi-targets are mature, we can change our way of thinking to review the issue of high-density multi-target tracking. This means not to distinguish those high-density multi-targets that are hard to identify when the conditions are not good; instead, track them as a whole and then conduct single-target tracking when there are conditions to separate the targets. In that case, high-density multi-target tracking translates into another issue—ensemble tracking of high-density multi-target cluster that is hard to identify. This ensemble tracking algorithm is feasible, so long as it is unnecessary to consider too much the carrying capacity of hardware, and the algorithm itself is simpler than that trying to distinguish every target and meets the mission require ment of radar. It is known that for any radar, the distance between targets always has a critical value which enables radar to reliably and stably track these targets, so as to avoid mistaken or missing tracking [16, 22]. With the changes of relative locations between radar and targets, some members will have larger or smaller distance than the critical value. Therefore, it is a transient process to form high-density multi-targets, either from high-density formation or from multiple bombs or missiles launched simultaneously by a weapon platform. Accordingly, high-density multi-targets can always be divided into identifiable single targets, unidentifiable high-density multi-target clusters, or both. That is to say, a high-density multi-target cluster might not be a high-density target to different

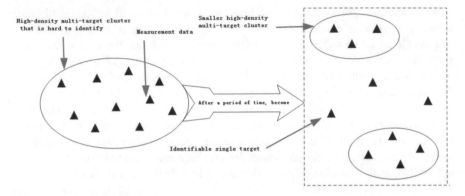

Fig. 1.4 Demonstration of high-density multi-targets changing into sparse multi-targets

radars. Therefore, we perform either single-target tracking for targets meeting correct association requirement, or ensemble tracking for high-density multi-target cluster, or tracking of single targets and target group simultaneously. As the above tracking method can change high-density multi-target tracking into sparse multi-target tracking, and sparse multi-target tracking technology is mature. Therefore, if we develop high-density multi-target cluster tracking algorithms, it means we can solve the problem of high-density multi-target tracking. Accordingly, high-density multi-targets are changed into "sparse" multi-targets, as shown in Fig. 1.4.

(3) Confirmation of group-target-tracking theory

For radar, multi-returns appear in conventional narrow-band radar and broadband radar, as well as single-target radar and multi-target radar. For target, multi-returns happen in STT and MTT, as well as appear in the multi-targets in the form of dot target and single extended targets. As regards the source of return signals from the target, there are real multi-returns from the target and the so-called indirect multi-returns from the target that are produced after secondary reflection, as well as unwanted false returns.

Now, we can draw the following conclusions.

Conclusion 1: Both single-target radar and multi-target radar involve multi-returns. That is, multi-target radar might not be able to totally solve multi-target tracking, and single-target radar might also face multi-returns.
Conclusion 2: Multi-targets naturally produce multi-returns, but the number of returns does not completely correspond to the number of targets.
Conclusion 3: Multi-returns of single target appear both in broadband radar and in extended targets of narrow-band radar.
Conclusion 4: Indirect return is neither a real return nor a false return, but a return that is of importance to target tracking.

Conclusion 5: Tracking of any target by any radar can be abstracted as the topic of multi-target tracking, that is, target tracking.

If we break through the boundary between single target and multi-target, between single sensor and multi-sensor, as well as the limitation that one target can correspond only to one real return in traditional tracking theory, we can expand the target tracking in a narrow sense to target tracking in a general sense. This is what we call the group-target tracking theory.

1.3.3   Group-Target Concept and Connotation

There are many names for formation target, such as target cluster [1–6, 8–18]. In essence, however, it refers to this type of targets: a group of multiple targets that are close to each other and fly almost along parallel tracks. Apparently, this definition is made based on the features of spatial location of targets. It has two implications. First, members of the formation target are close to each other, meaning close spatial location between targets. Second, the track of every target is almost parallel, meaning almost same speed and moving direction. Due to such demanding restrictions, formation targets can only be a particular case in high-density multi-targets. Typical high-density multi-targets include those multi-targets formed by the plane and the missiles it fires after they separate from each other. However, except for close target distance, the speed and moving direction of the plane and the missiles are not qualified for formation target. Therefore, formation target algorithm is not applicable to all scenarios of high-density multi-target tracking. If we relax the restriction on spatial location and motion features, and add some new limits, it will expand the applicable scope of radar for tracking such targets, and this will greatly enrich the contents and significance of relevant researches.

The group-target concept is defined based on two basic considerations. First, based on existing target tracking theories, change "high density" to "sparsity," that is, change high-density multi-targets to unidentifiable group-target and identifiable multi-target. Second, regardless of the quantity and density of targets, define it as a group-target even there is only one target. In theory, these two ideas can address tracking of multi-targets in high density, but the latter has apparently advanced from the data layer in traditional target tracking to situation perception layer. As a result, the group-target now is used not only to address the high density and sparsity of multi-targets, but to constitute what kind of group-target based on actual needs. It just depends on the criteria formulated.

The next is the group-target concept in a spatial sense based on the second consideration. It is defined as the assembly of multiple targets that maintain relatively stable spatial location for an adequately long period of time based on a given criteria for target distance. Single target and formation target are two particular cases of group-target. This concept has four implications. (i) Given target distance, that is, specific standard and formula are provided and qualitative statements like "the distance between two targets is very close" are no longer used. (ii) The condition for an adequately long period of time allows for certain speed difference between members, which therefore reduces the limitations on speed and removes the demanding requirement for members of formation target to fly in parallel tracks. (iii) Relatively fixed spatial location specifies the relative spatial location relationship between members of the group-target. (iv) The number of targets is no longer set.

While the term "group-target" has appeared in some publications, it is substantially different from the concept in this book. The so-called group-target in some publications is same as names like formation target, target group and target cluster in connotation and algorithm, and does not study the tracking of group-target as an independent equivalent target. On the contrary, the group-target and its tracking algorithm are measurement data and the corresponding tracking algorithm totally involving data association. Group-target tracking and formation target tracking are different most remarkably in the fact that the equivalent measurement data of the former contain the integrated virtual measurement data of all members, while that of the latter are single actual measurement data obtained by irradiating the geometric center of the formation target.

In short, the connotation of group-target can be summarized as follows: (i) classifying high-density multi-targets into unidentifiable high-density multi-target cluster and identifiable single targets; (ii) processing unidentifiable high-density multi-target cluster as group-target according to some rules; and (iii) tracking group-target and identifiable single targets as sparse multi-targets.

1.3.4   Group-Target Tracking Theory

(1) Fundamental principles

The mature theories and information processing architectures for multi-target and formation target tracking provide important references for group-target tracking. The following group-target tracking architecture is built based on the components of formation target and multi-target tracking and information processing architecture, as shown in Fig. 1.5.

Compared with multi-target tracking architecture, group-target tracking architecture is added with two function units—grouping detection and combination/separation detection. The latter function unit is equivalent to its data preprocessing and formation function unit and merges maneuvering base and judgment function unit. Compared to formation target tracking architecture, formation target detector is equivalent to grouping detection function unit of group-target tracking. The two are similarly located at the input position of the formation target tracking function unit. Separation target detector is equivalent to the combination/separation detection function unit of group-target tracking and is located at the output position of the formation target tracking function unit. But

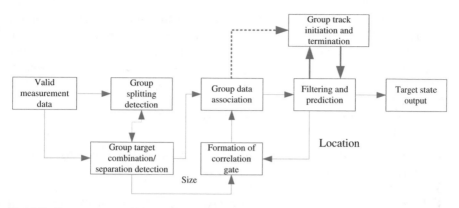

Fig. 1.5 Group-target tracking architecture

group-target tracking architecture moves the combination/separation detection function unit from output position to input position. Meanwhile, the group initiation function is added. That is to say, the information processing architecture for tracking of formation targets processes single target and formation target as two loops, while that for tracking group-targets processes two loops as one loop, like multi-target tracking architecture, which fully reflects the idea of group-target tracking.

(2)   Main problems to be solved by group-target tracking

Group-target is an all-new concept developed on the basis of formation target concept. Formation target tracking is a high-density multi-target tracking method without the need to know every target track. Group-target tracking is a high-density multi-target tracking method when it is difficult to distinguish every target and also a single-target tracking and formation target tracking method. While group-target tracking is similar to traditional multi-target tracking in terms of concept, process, and algorithm, it needs to add such functions as formation of group-targets as well as group-target combination/separation detection. In addition, due to the difference in boundary conditions and application scenarios, group-target tracking has new characteristics in algorithm.

Based on the basic architecture of group-target tracking, the major problem group-target tracking needs to solve is the specific algorithm enabled by each function unit. This mainly includes grouping detection and group initiation, single-group-target data association and track maintenance, multi-group-target data association and track maintenance, group-target combination detection, group-target separation detection, and group termination, among other algorithms. Group-target data association involves target tracking at the data layer and also situation association at the situation perception layer. Group-target combination/separation detection involves maneuver detection in target tracking, as well as situation tailoring, situation integration, and situation fusion. Therefore, it is a totally different concept and algorithm from traditional tracking methods.

It is worth special noting that the separation detection algorithm in group-target tracking is easily mixed with the method of low-resolution radar for distinguishing the number of formation targets. The former enables detection of separated targets at data layer based on such information as spatial location, direction, and speed of targets tracked, while the latter enables distinguishing of targets at signal layer based on the time–frequency characteristics of the signal. At present, the most commonly used methods are DBS and WT (wavelet transform), etc. And the two methods shall not be mixed.

(3) Problems in the application of formation target tracking algorithm in group-target tracking

(i) Formation target tracking algorithm cannot completely solve tracking of targets in high density

Formation target tracking is a multi-target tracking method for targets with same motion features, without the need to give every target track. However, this method is not completely applicable to tracking of high-density multi-targets with different motion features.

(ii) Grouping

Observation data-based group splitting method uses serial judgment to classify every target based on every measurement data, and more targets need more time. Track-based group splitting method adds density threshold restriction, but does not describe the relationship between target distance and track density.

The above two methods mention distance and density of targets, but do not provide the basis for these threshold values. And this is the basic standard for group splitting in group-target tracking algorithm.

(iii) Track initiation

Similarity method and furcation method are effective without considering interference and false alarms. When there are interference and false alarms, however, group initiation will be quite difficult. That is to say, as a single measurement data initiation method, this method is only applicable to strict formation group-target, but not to the group-target track initiation defined in this book.

K-means has the following main defects [17]. (i) The number of cluster members is needed as an initiation condition; otherwise, this process will go on without end until there is no target in the surveillance area. (ii) The convergence process is slow. (iii) Noises or noise waves will impact the correctness of cluster initiation. The requirement for the number of targets is very demanding, as factors like false alarm will make it almost impossible to know the number of targets. As a result, it is necessary to use the auxiliary diagrams and other means to obtain the K value, but this is time-consuming and thus is not very practical. The weakness of cluster-seeding track initiation method [17] is that the sequence of returns will impact formation of cluster, which may lead to wrong results. We can see some improvements in this method, such as checking the distance between clusters and threshold value, and if it is smaller than the threshold value, clusters will be combined into one cluster. However, it is still time-consuming even after being improved. Though K-means and cluster-seeding track initiation method have such advantages as simultaneous group splitting and group initiation, they are time-consuming which leads to poor practicality. Therefore, they will not be analyzed in detail in the book.

(iv)   Data association method

Formation target tracking algorithm neither considers formation target as an independent "target," nor has its own tracking gate and corresponding data association algorithm. Rather, it adopts traditional multi-target data association algorithm. Therefore, group-target tracking algorithm needs to develop new relevant threshold and data association algorithms.

(v)   Combination and separation of formation targets

Combination of formation targets is enabled based on the judgment on whether the minimal value of observation data distance of the two subgroups is smaller than the specified target distance. This method is only applicable to the case that the tracks of the two subgroups are totally parallel.

Separation of members of a formation target is based on the assumption that only the targets on the margins of the formation will separate. Thus, it uses the method to track the targets on the margins of the formation. That is to say, while tracking the center of the formation target, track targets on the margins of the formation with two beams and checks whether to combine and separate formation members based on specified target distance criteria. However, when formation target is coming from the front or tail of the radar, this method can only help detect the two margins of the formation and cannot detect the separation occurrence in the front as well as above and under the formation, especially separation under the formation that is more likely to happen. When a formation target is at the approach point of the radar, the radar can only detect the separation of the members in the front, as it is almost impossible for formation members to leave the formation from the rear. Therefore, this method has inherent limitations.

(vi)   High requirement for tracking environment

Formation target tracking does not consider the impact of false measurement data like noise waves [2, 8–13] and is applicable to tracking of targets in clean environment, such as in medium and high altitudes, which thus limits its application in group-target tracking.

1.3.5   Application Prospect of Group-Target Tracking

(1)   It enables unification of target tracking systems with different types of sensor

Traditional target tracking systems involve STT by single sensor, MTT by single sensor, STT by multi-sensors, and MTT by multi-sensor. Moreover, multi-targets include multi-targets in sparsity and multi-targets in high density. As a result, when designing each sensor, we have to consider its specific target, which leads to multiple types of sensors for adaptability to various kinds of targets. This is harmful to standardized, universalized, and modular design of sensor systems. Group-target tracking enables unification of STT and MTT, unification of MTT in sparsity, and MTT in high density. Actually, all tracking systems with all types of sensors can be unified.

(2)   It has the potential to make threat judgment and quickly capture threat targets

It is of important significance in situation perception in the battlefield to timely discover and track targets with invasion tendency and behavior. For instance, radar is tracking a two-plane formation by means of group-target tracking. At a certain moment, each of the two planes fires a missile at the same time, and instantly, the number of group-target members increases from two to four. However, the radar cannot immediately distinguish returns from the two missiles. Then, the missiles leave the formation rapidly after engine ignition to attack against preset targets. In this case, the radar needs to timely capture and track the missiles while continuously tracking the plane formation.

(3)   It has the ability for comprehensive situation perception in the complex battlefield environment

The basic conception is as follows. In a vast battlefield, there are many targets of enemy, friend, and our own forces with different spatial locations, identity attributes, extent of threat, etc. We classify these targets into different group-targets for tracking and management with identifying information provided by identifiers of the enemy and our side. For example, treat targets with the biggest threat as a group, targets away from the battlefield as a group, targets hard to distinguish whether being friend or enemy as a group, etc.

(4)   Research on group-target imaging and its tracking algorithm

Based on the characteristics of relatively fixed spatial location between members of a group-target, this method considers multiple returns in the association gate as the returns of high-resolution radar. It enables group-target imaging according to the principle for broadband radar imaging and then conducts real-time surveillance of the overall moving situation of the group-target through the images of the target group. On the basis of TV tracking theory, it enables tracking of group-target by treating the association gate of the group-target as the scene for TV tracking system, the tracking gate as TV tracking gate, the predicted center of equivalent measurement data as reticule, and equivalent measurement residual as miss distance. Such imaging method needs to consider the special problems caused by a group-target as non-rigid body and find ways to correct them.

(5) Integration of broadband radar imaging and tracking

At present, some radars add a narrow pulse signal before imaging broad pulse and use narrow pulse for tracking of target and broad pulse for imaging. This method is easy to use, but has some disadvantages. Narrow pulse tracking cannot guarantee radar beams always point at the center of the target in addition to wasting of precious radar resources. Group tracking method not only removes narrow pulse to save radar resources, but can guarantee radar beams stably point at the center of the target.

(6) It has the ability of centroid tracking by terminal guidance radar

For a terminal guidance radar, a target becomes extended targets as it approaches. Extended targets will emit multi-returns, causing increased guidance miss errors, even off-target. Group-target tracking can solve this problem.

(7) It can track space debris and microsatellite formation

Space debris, or space trash, can harm the safety of spacecraft, which has been drawing increasing attention of all countries in the world. As space debris and microsatellite formation have relatively stable position relation, using group tracking algorithm to track space debris and microsatellite formation is of great significance to saving space target detecting resources.

(8) It can be used to identify the quantity of formation targets

At present, such methods as DBS and WT enable low-resolution radar to identify the quantity of formation targets. If the method to distinguish targets at this signal layer is combined with the separation detection method at the data layer of the group-target, the distinguishing effect will be greatly enhanced.

(9) It has certain anti-interference capability

Either active or passive interference features increased the number of returns and density. Therefore, in the case that it is inconvenient or impossible to track all targets one by one, group-target tracking can be used to comprehensively monitor and timely track the targets separated from the formation. In this sense, group-target tracking has certain anti-interference capability.

1.4   Scope and Profile

This book highlights group-target concept, basic theory and concept for group-target tracking, tracking method and relevant technologies, etc. It mainly involves the formation of group-target based on observation data, group-target track initiation, group-target association algorithm, group-target combination/separation detection, and group-target track termination. With seven chapters, its main contents can be summarized as "one concept, two cores, and three keys," that is, the concept of group-target tracking; single-group-target data association algorithm and multi-group-target data association algorithm; and algorithm for group splitting/combination/separation detection of group-targets.

This chapter is introduction, which generalizes the development history of target tracking and typical target tracking systems, defines group-target concept and connotation, establishes group-target tracking architecture, and elaborates the formation process of group-target tracking idea, basic principles, and application prospects. It is aimed to raise questions and provide an overall description of group-target.

Chapter 2 is about basic theory, which introduces commonly used terms, Kalman filtering, target motion model, and basic algorithms for target tracking. It is aimed to provide basis knowledge for research on group-target tracking algorithms.

Chapter 3 is about group splitting detection of group-target and group initiation, which provides criteria for determining the distance in group-target based on the analysis that group splitting method for formation target has problems. It also elaborates the algorithm for group splitting detection of group-target. And it goes on to introduce the track initiation algorithm based on the geometric center of the group-target. This part constitutes the foundation for enabling group-target tracking.

Chapter 4 is about single-group-target data association, which mainly describes single-group-target association gate and tracking gate association method and elaborates the method for calculating measurement weight based on the nearest-neighboring and all-neighboring concept as well as single-group-target data association and track maintenance algorithm.

Chapter 5 is about multi-target data association, which mainly introduces the basic idea that breaks through the restrictions of traditional algorithm requiring one-to-one correspondence of real measurement data to target, establishes the matrix for cross-detection of association gates of multi-targets, and expatiates on the method for calculation of equivalent measurement data of multi-group-targets under double many-to-many correspondence.

Chapter 6 is about combination and separation detection of group-target and situation perception, which analyzes the problems in separation detection of formation target, provides basic description on the maneuvering of group-target, and elaborates group-target combination and separation algorithm as well as smooth transition of group-target centroid and group-target track termination.

Chapter 7 is about group-target algorithm simulation, which, based on given simulation conditions and simulation platform, makes simulation of data association between single-group-target and multi-group-targets, algorithm for detection of group-targets through combination and separation, and other algorithms. The simulation results prove the correctness and effectiveness of the group-target algorithm.

1.5   Summary

This part briefly reviews the development history of target tracking, describes the forms of multi-targets, elaborates basic issues related to tracking of multi-targets in high density, and summarizes problems in tracking of multi-targets in high density. In addition, it also introduces typical tracking systems and analyzes in detail special cases of high-density multi-targets—research status of formation target tracking, major technologies used, and problems during application of these technologies in tracking of group-targets.

Drawing inspirations from the concept for ensemble tracking of formation target as well as multi-target tracking theories and referring to the components of multi-target tracking algorithms, it defines group-target concept and elaborates the background, basic idea, and basic principles of group-target tracking. Based on this, it builds the basic architecture for group-target tracking and analyzes the application prospects of group-target tracking. The research of group-target tracking focuses mainly on group splitting detection and group initiation, single-/multi-target data association and track maintenance algorithm, detection of group-targets through combination and separation, and group termination algorithm, among others. Compared to multi-target tracking process, group-target tracking algorithm adds two function units—group splitting detection and group initiation, and target combination/separation detection. Compared to formation target tracking process, it adds group splitting detection and group initiation and moves target combination/separation detection from input position to output position.

Chapter 2
Preliminaries

Modern radar systems include radar signal processors, data collectors, and radar data processors, besides traditional antennas, transmitters, receivers, and monitors. Signal processors are used mainly for target signal detection and for restraining the irrelevant signals of noise wave disturbances from earth or ocean, multi-path effects, atmospheric environment, universal noises, and man-made disturbances. The video output signals after signal processing are compared with certain detection threshold. If the signals are beyond the detection threshold, we "discover" targets. Then, we send the signals to data collectors, to measure the distance, angle, and radial velocity of the target, and for some radar, we can even measure the dispersion, polarization, and geometry of the target. The output of data collectors is the approximation of target observation. The observation from tradition radar is called observation data or measurement data. The observation data output by data collectors will be dealt with by data processors for all kinds of relevant processing. Radar data processing is used for computations such as association, tracking, filtering, smoothing, and prediction, after radar gets the location and motional parameters such as radial distance, radial velocity, and azimuth and pitch angle of targets. The processing of radar measurement data can effectively restrain the random errors introduced during the target measuring, precisely estimate the parameters relevant to location and motion of targets, and get stable tracks of targets.

From the perspective of information processing, if radar signal processing is the first-time processing, radar data processing is the second-time processing, and situation evaluation and threat judgment is the third-time processing. Some references called the interception determination, interception instruction computation, and interception means selection and kill probability computation the third-time processing. And some references in the early Soviet Union called data fusion the third-time processing. In the recent years, with the development of hardware, algorithms, and computer performance, there is qualitative leap in the information processing capability of radar, which enables the connectivity between "the physical, data, and cognitive layers" in radar. It is why the authors propose the group-target tracking algorithm.

© National Defense Industry Press and Springer Science+Business Media Singapore 2017   29
W. Geng et al., *Group-target Tracking*, DOI 10.1007/978-981-10-1888-6_2

Target tracking is an interdisciplinary branch of science integrating information processing technology, cybernetics, and modern mathematics. There are many foundational theories, and we cannot elaborate in detail. This chapter will only expatiate on the content relevant to the research of this book.

2.1 Common Glossary

Next, we concisely introduce the basic glossary relevant to the target tracking. Please note that what we give in the following is glossary for multi-target, and the glossary for group-target will be given in the remainder chapters.

(1) Measurements

Measurements, also called measures, observations, or observation data, are the measurement data relevant to the state of targets and are contaminated by noises. Modern radar measurement is unnecessarily the original data output by receivers, but is the observation data output by data collectors after signal processing. The types of measurement include target radial distance, azimuth and pitch angle, radial velocity, the differences in time, frequency, and phase between the arrivals of signal to two sensors, and signal frequency and range transmit by targets.

Measurements come from possibly correct measurements of targets, false measurements of noise waves from earth/ocean, cloud, and rain, mistaken measurements of unexpected targets and baits in the target environment, and also false-alarm signal due to radar noises. Furthermore, missing detection of target signals may occur. So there are uncertainties in measurements.

(2) Tracks

The estimation of target state, according to the set of measurements from one target, goes through computation to give the track of the continuous motion of targets. A complete track also includes a track number, track quality, possible tracks, tentative tracks, validated tracks, terminated tracks, and response time of track initiation. The number of track can be regarded as the reference to all parameters related to the track. Track quality indicates the reliability of tracks, through which we can timely and correctly initiate tracks to build up new target archives and can also timely and correctly terminate tracks to remove redundant target' archives. Possible tracks consist of single measurement spots. Tentative tracks consist of two or more observation data and have comparatively low track quality. They can be target tracks or false tracks caused by random disturbances. Tracks, after initiation, will be transformed into tentative tracks or terminated tracks, so tentative tracks are also called temporary tracks. Validated tracks are those that have stable outputs or track quality of some standard. They are also called reliable tracks or stable tracks and usually are seen as real target tracks. Terminated tracks are those that have track quality below some fixed value or consist of isolated observation data. And the corresponding process is called track termination or track ending. The main task for track termination is to timely delete false tracks and keep real tracks. Response time of track initiation is the time period from the entering of target into radar power area to the construction of the track, usually taking radar scan time as its unit. Quick track initiation usually takes 3–4 radar scan cycles, while low track initiation usually takes 8–10 radar scan cycles.

(3)   Data association

In a general sense, data association means the buildup the connection between the measurement data at one time and the measurement data or tracks at other time. It is the procedure to determine whether there measurement data come from the same target, or to pair correct observation data and tracks.

The objects of data association include track initiation, track maintenance, and track association. The association between measurement and measurement, or observation data and observation data, is the procedure of generating stable tracking according to the observation data obtained at some time. Track maintenance, also called track updates or track preservation, means the association between measurement and track, or observation data and track. Track association, also called track integration, means the association between track and track. Mathematically, data association can be divided into deterministic model and probability model. For the deterministic model, the source of measurement is deterministic, and the truth that it is unnecessarily correct is ignored. For the probability model, according to the occurrence probability of each true/false event, correct the approximation of target state through the probability. They two are corresponding to different data association algorithms, which will be discussed in detail in the following sections.

(4)   Tracking

Data association and tracking are the two basic issues in radar data processing. There is only tracking issue for single targets, while there are both association and tracking issues for multi-targets.

Tracking means the processing of measurements derived from targets, to maintain the approximation of the current state of targets. The typical state of targets includes dimension information such as location, velocity, and acceleration; "signal characteristic" information such as strength, frequency, pulse width, and pulse recurrence frequency of transmit signals; and constants or slowly varying parameters such as coupling coefficient and transmission velocity of electromagnetic wave or sound waves.

Multi-target tracking means the concurrent processing of measurements from multi-targets, to maintain the approximation of the current state of multi-targets.

(5)   Association gates

Association gates, also called association gates of tracking gates, take the predicted location of target under tracking as center and determine the area where the measurements of this target possibly occur. The dimension of the area is subject to the probability of correctly receiving target returns. Association gates have shapes such cuboid, cube, sphere, and ellipsoid. The principle of determining the shape and dimension of association gates is to make true measurements fall inside association gates most likely and in the same time make the number of irrelevant observation data as little as possible.

(6) Motional models

Motional models are the assumptions about the motional rules of targets. We can only obtain the state equations under some assumptions. Maneuver represents the uncertain changes in target motion, such as sudden acceleration and turning, which is one of the difficulties in target tracking research.

(7) State approximation

State approximation is the smoothing of the past motional state of targets, such as location, velocity, and acceleration, the filtering of the current motional state, and the prediction of the future motional state of targets.

2.2 Kalman Filtering

Literally, filtering is to filter waves. For example, in an automatic control system, the automatic control of system is realized through the feedback of system outputs. The system outputs usually include some disturbing signals and noises. When distilling some quantity of feedback as control quantity, there always exist random errors. Thus, we must appropriately filter in order to decrease control errors.

From the prospective of classical filtering, useful signals and noises are distributed within different frequency bands (sometimes overlaps possibly happen). Thus, we can use some classical filtering networks with certain frequency selection features to eliminate noises as many as possible, and keep useful signals with little aberration. However, the signals and noises we meet with sometimes are random, and their characteristics can be only described statistically. For example, the flying motion of missiles is random. The location and velocity are naturally random; meanwhile, measuring equipment has random errors as well. Therefore, we cannot adopt general classical filters to separate useful signals from measuring results, but can only compute the optimal estimation of useful signals through statistically estimation methods. From the statistical prospective, the closer the outputs of a filter are to practically useful signals, the better the performance of the filter is.

During the course of statistical filter development, early Wiener filter performs static processing of the time-independent statistical characteristics of signals. In this procedure, the statistical characteristics of useful signals and useless noises can be connected with their frequency-domain characteristics. Thus, Wiener filter and classical filters come down in one continuous line in conception. Wiener filter is applied widely during the Second World War. However, the demerits of Wiener filter are as follows: (1) The all-historical observation data must be used, which leads to great storage and computation cost; (2) when new observation data are obtained, we cannot perform recursion but have to recalculate; and (3) it can hardly be applied to filtering for unstable processes.

To overcome the demerits of Wiener filter, Kalman proposed a filtering method with recursion function in the early 60s, which is called Kalman filtering. Different from Wiener filter, Kalman filtering processes time-varying statistical characteristics, but not consider from the angles of frequency domain or time domain. In the last half-century, Kalman filtering has been widely applied in many domains.

2.2.1   Filter Description

According to the diverse principles in parameter estimation, there are four basic methods including maximum likelihood estimation, maximum a posterior estimation, least square estimation, and minimum mean square error estimation. All these estimation methods are used for time-independent parameters. In the nature, many parameters are a time function, which leads to the estimation problem of time-varying parameters. The estimation of time-varying parameters is also called state estimation. Since unknown parameters are a time function in state estimation, we must consider the evolvement of unknown parameters and observation data with time, during the course of processing measurement data.

State variable method is a valuable method for describing dynamic systems. Using this method, the input and output relationships of systems can be discussed in time domain through the state transition model and the output observation model. Inputs can be described by dynamic models consisting of certain time functions and stochastic processes representing unpredictable variables or noises. Outputs are a function of states, usually disturbed by random observation errors, and can be described by measurement equations. For physical clarity, we use both figures and mathematical expressions to describe the process.

As shown in Fig. 2.1, the state vector is denoted by $X(k)$, the measurement vector is denoted by $Y(k)$, and $Z^{-1}I$ represents storage unit. The mathematical description of state equation and measurement equation is as follows:

(1) State equation

$$X(k+1) = \Phi(k+1,k)X(k) + G(k)W(k) + U(k) \tag{2.1}$$

where $X(k) \in R^{n\times 1}$ is target state vector, $W(k) \in R^{p\times 1}$ is known inputs or control signals, $U(k) \in R^{n\times 1}$ is the white Gaussian noise sequence with zero mean, $Q(k)$ is covariance matrix, and $\Phi(k+1,k) \in R^{n\times n}$ and $G(k) \in R^{n\times p}$ are state transition

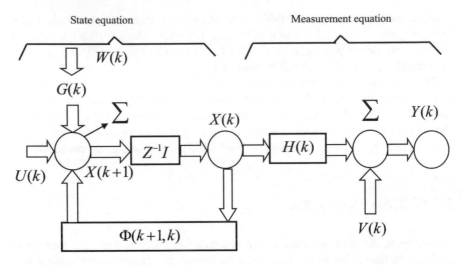

Fig. 2.1 The information flowchart of linearly dynamic discrete systems

matrix and input matrix, respectively. The association function of $U(k)$, white Gaussian noise, is:

$$E\big(U(i)U^T(k)\big) = \begin{cases} Q(k), & i = k \\ 0, & i \neq k \end{cases} \tag{2.2}$$

$\Phi(k+1,k)$, as the transition matrix from time k to time $k+1$, has the following characters.

(a) Multiplication law

$$\Phi(n,k)\Phi(k,m) = \Phi(n,m) \tag{2.3}$$

(b) Inversion law

$$\Phi(n,m)^{-1} = \Phi(n,m) \tag{2.4}$$

where k,m,n are integers.

(2) Measurement equation

$$Y(k) = H(k)X(k) + V(k) \tag{2.5}$$

$Y(k) \in R^{m \times 1}$ is measurement vector, $V(k) \in R^{m \times 1}$ are irrelevant measurement noises, and its covariance matrix is $R(k)$. $H(k) \in R^{m \times n}$ is observation matrix.

The association function of $V(k)$, white noises with zero mean, is:

$$E\left(V(i)V^T(k)\right) = \begin{cases} R(k), & i = k \\ 0, & i \neq k \end{cases} \tag{2.6}$$

Process noise and observation noise are independent from each other. So we get:

$$E\left(U(i)V^T(k)\right) = 0 \quad \forall\, i, k \tag{2.7}$$

The measurement as in Eq. 2.5 identifies the relationship between measurable system output $Y(k)$ and state $X(k)$. The unknown state can be computed by arraying the state equation and observation equation, with a predefined optimization manner. Also, we assume that the initial state $X(0)$ is a Gaussian sequence.

In short, there is three prior information for implementing Kalman filtering discrete-time dynamic systems. First, initial state $X(0)$ is a Gaussian sequence. Second, initial state is independent to process noises and measurement noises. Third, process noises and measurement noises are independent from each other. On this condition, the linear character of state equation and measurement equation can keep the Gaussian character of state and measurement. According to the known measurements at time i and before time i, some kind of estimation of state $X(k)$ at time k is denoted by $\widehat{X}(k, i)$. According to the time indicated by state estimation, Kalman filtering can be described as follows.

When $k = i$, it is called filtering, and $\widehat{X}(k, i)$ is the filtered value of state $X(k)$ at time k;
When $k < i$, it is called smoothing, and $\widehat{X}(k, i)$ is the smoothed value of state $X(k)$ at time k;
When $k > i$, it is called prediction, and $\widehat{X}(k, i)$ is the predicted value of state $X(k)$ at time k.

2.2.2   Filtering Models

The Kalman filtering, under the law of linear mean square error estimation, is very suitable for recursive algorithms to smooth the past and current state of targets and to predict the future motional state of targets, including parameters such as, velocity and acceleration.

The least mean square error estimation of state vector is defined as follows:

$$\widehat{X}(k/k) = E[X(k)/Y_k] \tag{2.8}$$

where $Y_k = \{y(k), k = 1, 2, 3, \ldots, m\}$. The covariance matrix of state errors companying with Eq. 2.8 is:

$$P(k/k) = E\left[\left(X(k) - \widehat{X}(k/k)\right) \cdot \left(X(k) - \widehat{X}^T(k/k)\right)/Y_k\right]$$
$$= E\left[\widetilde{X}(k/k) \cdot \widetilde{X}^T(k/k)/Y_k\right] \tag{2.9}$$

where $\widetilde{X}(k/k)$ is the filtering error. If the error joints with the state equation, we can get the one-step prediction of state:

$$\widehat{X}(k+1/k) = E[X(k+1)/Y(k)]$$
$$= E[\Phi(k+1,k)X(k) + G(k)W(k) + U(k)/Y(k)]$$
$$= \Phi(k+1,k)\widehat{X}(k/k) + G(k)W(k)$$

The error of the one-step predicted value is:

$$\widetilde{X}(k+1/k) = X(k+1) - \widehat{X}(k+1/k) = \Phi(k+1,k)\widetilde{X}(k/k) + U(k) \tag{2.10}$$

The covariance for computing the error of the one-step predicted value is:

$$P(k+1/k) = E\left[\widetilde{X}(k+1) \cdot \left(X(k+1) - \widehat{X}^T(k+1/k)\right)/Y_k\right]$$
$$= \Phi(k)P(k/k)\Phi^T(k) + Q(k) \tag{2.11}$$

Please note that the covariance for computing the error of the one-step predicted value is a symmetric matrix and can be used to measure the uncertainty of predictions. The smaller the $P(k+1/k)$ is, the more precise the prediction is.

Taking the expectation of measurement equation at time $k+1$ under the condition of Y_k, we can similarly get the one-step prediction of measurements:

$$\widehat{Y}(k+1/k) = E[Y(k+1)/Y_k] = H(k+1)\widehat{X}(k+1/k) \tag{2.12}$$

The error of the one-step predicted measurement is:

$$v(k+1) = \widetilde{Y}(k+1/k) = Y(k+1) - \widehat{Y}(k+1/k) = H(k+1)\widetilde{X}(k+1/k) + U(k) \tag{2.13}$$

As this is a measurement of the new information generated during the process of prediction, the prediction error is called "new information" and is also called measurement residual in some references.

The covariance of the new information is:

$$S(k+1) = E\left[\widetilde{Y}(k+1/k) \cdot \widetilde{Y}^T(k+1/k)/Y_k\right]$$
$$= H(k+1)P(k+1/k)H^T(k+1) + R(k+1) \tag{2.14}$$

The covariance of the new information is also a symmetric matrix and can be used to measure the uncertainty of the new information. The smaller the covariance is, the more precise the measurement value is.

The covariance between state and measurement is:

$$P_{XY}(k+1/k) = E\left[(\widetilde{X}(k+1/k) \cdot \widetilde{Y}^T(k+1/k))/Y_k\right] = P(k+1/k)H^T(k+1)$$

$$(2.15)$$

Filter gains are:

$$K(k+1) = P_{XY}(k+1/k)S^{-1}(k+1) = P(k+1/k)H^T(k+1)S^{-1}(k+1) \quad (2.16)$$

So the equation for computing the estimation update of state at time $k+1$ is:

$$\widehat{X}(k+1/k+1) = \widehat{X}(k+1/k) + K(k+1)v(k+1) \qquad (2.17)$$

The equation tells that the estimation $\widehat{X}(k+1/k+1)$ at time $k+1$ is the sum of the predicted value $\widehat{X}(k+1/k)$ at this time and a correction item. And this correction item is relevant to gains and the new information.

Furthermore, the updating equation of covariance is:

$$
\begin{aligned}
P(k+1/k+1) &= P(k+1/k) - P(k+1/k)H^T(k+1)S(k+1)H(k+1)P(k+1/k) \\
&= [I - K(k+1)H(k+1)]P(k+1/k) \\
&= P(k+1/k) - K(k+1)S(k+1)K^T(k+1) \\
&= [I - K(k+1)H(k+1)]P(k+1/k)[I - K(k+1)H(k+1)]^T \\
&\quad - K(k+1)R(k+1)K^T(k+1)
\end{aligned}
$$

$$(2.18)$$

where I is the unit matrix with the same dimension as the covariance matrix. Equation 2.18 can guarantee the symmetry and positive definiteness of the covariance matrix P.

Another expression for filtering gain is:

$$
\begin{aligned}
K(k+1) &= P(k+1/k+1)H^T(k+1)R^{-1}(k+1) \\
&= \big[P(k+1/k)H^T(k+1) - P(k+1/k)H^T(k+1)S^{-1}(k+1) \\
&\quad H(k+1)P(k+1/k)H^T(k+1)\big]R^{-1}(k+1) \\
&= P(k+1/k+1)H^T(k+1)S^{-1}(k+1) \\
&\quad \big[S(k+1) - H(k+1)P(k+1/k)H^T(k+1)\big]R^{-1}(k+1)
\end{aligned}
$$

$$(2.19)$$

2.2.3   Summary of Filtering and Prediction Models

According to the above-mentioned analysis, the filtering basic equation and one-step prediction equation are as follows, respectively.

Filtering estimation equation is:

$$\widehat{X}(k/k) = \widehat{X}(k/k-1) + K(k) \cdot \left[Y(k) - H(k)\widehat{X}(k/k-1) \right] \qquad (2.20)$$

State prediction equation is:

$$\widehat{X}(k/k-1) = \Phi(k/k-1)\widehat{X}(k-1/k-1) \qquad (2.21)$$

The filtering gain matrix is:

$$K(k) = P(k/k-1)H^T(k) \cdot \left[H(k)P(k/k-1) \cdot H^T(k) + R(k) \right]^{-1} \qquad (2.22)$$

The covariance of prediction errors is:

$$\begin{aligned} P(k/k-1) &= \Phi(k/k-1)P(k-1/k-1) \cdot \Phi^T(k/k-1) \\ &\quad + G(k-1) \cdot Q(k-1)G^T(k-1) \end{aligned} \qquad (2.23)$$

The covariance of filtering estimation is:

$$P(k/k) = [I - K(k)H(k)]P(k/k-1) \qquad (2.24)$$

The predicted value of measurements is:

$$\widehat{Y}(k) = H(k)\widehat{X}(k/k-1) \qquad (2.25)$$

The new information (residual) vector is:

$$v(k) = Y(k) - H(k)\widehat{X}(k/k-1) \qquad (2.26)$$

The covariance matrix of residual vector is:

$$S(k) = H(k)P(k/k-1)H^T(k) + R(k) \qquad (2.27)$$

The one-step equation is:

$$\widehat{X}(k+1/k) = \Phi(k+1/k)\widehat{X}(k/k-1) + K(k)\left[Y(k) - H(k)\widehat{X}(k/k-1) \right] \quad (2.28)$$

The one-step predicted gain matrix is:

$$K(k) = \Phi(k+1/k)P(k/k-1) \cdot \left[H(k)P(k/k-1)H^T(k) + R(k) \right]^{-1} \quad (2.29)$$

The one-step covariance matrix is:

$$P(k+1/k) = [\Phi(k+1,k) - K(k)H(k)] \cdot P(k/k-1) + G(k)Q(k)G^T(k) \quad (2.30)$$

2.3 Target Motional Models

2.3.1 Problem Description

(1) Problem posing

Target tracking generally includes target maneuver model identification, maneuver detection, estimation of maneuver quantity, and tracking algorithms. When the target maneuver model is inconsistent with the practical motional situation, mistaken track association or loss of targets can happen. Thus, the target motional model is a key issue in target tracking domain.

The kinetic model of target motion can generally be categorized into two classes—evenly variable target model and maneuver target model. Evenly variable motion means even motion, evenly accclerated motion, and evenly decelerated motion. Maneuver means that there are sudden changes in direction and acceleration during the process of target motion. The statistical characteristics of systematical dynamical noises due to stochastic factors, such as atmospheric disturbance, and observation noises can be correctly described in advance. But maneuver is stochastic and sudden and the direction and quantity of maneuver is unknown, so that we can hardly build up the kinetic model of the maneuvering target timely and correctly. Kalman filtering has good performance in non-maneuvering target tracking, but bad precision in maneuvering target tracking, or even divergence of filters. Therefore, the key is to precisely build up a maneuvering target model to support the problem of maneuvering target tracking. This is why the maneuvering target model is emphasized all the time.

The maneuver model, as one of the basic elements in maneuvering target tracking, is a key but also a challenge, especially in aspects of maneuver detection of sudden changes in velocity and direction of targets and the estimation of maneuver quantity. The maneuvering target model needs to be consistent with the practical maneuver and to be easily dealt with mathematically.

The maneuver detection of target motion is a kind of judgment mechanism, to deal with the problem of the inconsistency between the existing motional model and practical situation upon the occurrence of target maneuver. This inconsistency essentially means the changes in the statistical characteristics of new information sequence caused by maneuver. Thus, the maneuvering target detector can be seen as a filter taking inputs from the new information sequence of Kalman filtering. Maneuver detection method mainly includes the method based on the statistical characteristics of new information sequence, the method based on the extrapolation errors, the geometric method based on the target motional tracks, and the methods of knowledge repository or neural networks emerging in recent years. No matter which method we adopt, we are faced with the following problems in practical applications:

First, maneuver detection threshold. The threshold value is subject to false-alarm probability and detection probability. Thus, maneuver detection confidence needs to be considered.

Second, the estimation lags of maneuver detection. The maneuvering target detection, as a method of a posterior processing, necessarily has lags. Thus, we need to choose an appropriate detection data window length, to timely track the practical maneuver situation of targets.

Third, the transient error problem during the transition between filters for maneuver and non-maneuver.

Fourth, the estimation of maneuver quantity. Currently, the least square method, self-adaptive algorithm, and the geometric method for detector and target inter-ception figure are mainly in use.

Currently, the comparatively effective method is the multi-model method, in maneuvering target tracking. But this method is faced with two difficulties. First one is the design of target maneuver characteristic detector. To guarantee that multiple models can completely cover all the possible maneuvers of targets, we need to comprehensively consider the influence of the signal-to-noise ratio change due to maneuver on the capability of maneuvering target detection, the influence of tracking errors due to maneuver on the capability of tracking, and the influence of filter-model parameter revising due to maneuver quantity on tracking precision. Second one is the robustness. On the one hand, maneuver detection needs to judge on the change in new information sequence mean. Whether it is caused by target maneuver or exterior disturbance, namely the source of abnormal values, needs to be determined. The difficulty for a tracking system lies in how to compromise between the judgment of abnormal values and the precision of system state esti-mation. Thus, the maneuver detection result without eliminating abnormal values is unreliable, before the processing of target maneuver detection. On the other hand, maneuver detection needs to judge on the loss of target observation data, due to target maneuver, system missing alarms, or exterior disturbances. The maneuvering target detection, under this circumstance, needs to comprehensively consider signal processing.

(2) Review of target maneuver models

In 1970, Singer regarded the acceleration caused by maneuver and atmospheric overfall as a disturbance to even motion and proposed a first-moment stationary Markov process to describe the maneuver acceleration. But this method is suitable for the situation when target has inconsiderable maneuver, but not for the situation of considerable maneuver. The important contribution of the Singer model lies in that it provides a new idea of treating the maneuver as a disturbance to system equations, to the research of maneuver models. In 1973, Mcaulay applied the statistical detection theory to maneuvering target Kalman filtering, constructed the maneuvering target detector, and for the first time comprehensively investigated the maneuvering target tracking problem. Mcaulay took advantage of the new information sequence characteristic of optimal filters, treated the maneuvering target detection problem as a binary hypothesis testing problem, and used two different target models to perform Kalman filtering under conditions of maneuver and non-maneuver, respectively. Thorp introduced a binary random variable to describe maneuver and non-maneuver, directly constructed two sets of Kalman filters, and denoted the final estimation value by weighted linear combination of the filtering results from the two sets.

In order to solve the problem of simply categorizing target motional state as maneuver and non-maneuver, Moose categorized the pattern of maneuvering targets into half-Markov process and Singer first-moment self-regressive process and constructed the correlated Gaussian noise model with random switch mean. He first divided maneuver acceleration into N possible maneuvering acceleration instructions, used a filter set consisting of N augmented Kalman filters to perform filtering, then computed the N possible filtered values of state at some time and the occurrence probability of each filtered value of state, and finally set the filtered value of state at this moment to be the weighted value of the probability. Although this method is more advanced compared to the Thorp method, it needs to choose N acceleration values in advance to approximate the acceleration value of the actual maneuvering target. Obviously, the complexity and computation cost increase with N. Chan regarded maneuver acceleration as an unknown constant and proposed a model to estimate the maneuvering target acceleration by using least square estimation. To degrade the complexity of the model, Chan further simplified the algorithm. Maneuvering detector was no more a detector based on new information sequence, but a detector based on the difference between the extrapolated filtering value and observation value. Bar-Shalom proposed a dimension-varying filtering method. A location–velocity 2-dimensional Kalman filter was adopted when targets had no maneuver, while a location–velocity–acceleration 3-dimensional augmented Kalman filter was adopted when targets had maneuver. In 1990, Bar-Shalom proposed an interacting multiple model tracking algorithm.

To improve the demerits of the existing maneuver models in expensive computational cost or computation delays, Wen proposed a type of increment model to perform detection and estimation of target maneuver characteristics, when using Kalman filter to track targets. The characteristic of the model is the introduction of a vector recording acceleration to the system equation. Take the known maneuver quantity at the last time point as a reference; assume the changes in acceleration to be a set of augmented maneuver; then predict the set of state estimation at the next time point by using the set of augmented values and reference values; and compute the mostly possible state estimation based on Bayesian equation. The mostly obvious feature of this method is the simplicity.

A Chinese researcher Zhou Hongren proposed a current statistical model of maneuvering targets. For each concrete tactical occasion, people care only the "current" probability density of maneuver acceleration, namely the current probability of target maneuver. When targets currently maneuver with acceleration, its value range of the acceleration in the next sampling time point is limited and can be within the neighboring range of the "current" acceleration. Therefore, when describing the probability density of maneuver acceleration, it is completely unnecessary to consider all the possibilities of maneuver acceleration values. Additionally, they are many other maneuvering target models. Here, we do not discuss in detail any more.

2.3.2 Basic Motional Model

According the above-mentioned analysis, no matter we use sampling-interval decreasing method or multiple models, they are essentially based on evenly variable motional models. Group-target tracking aims at obtaining the holistic equivalent measurement, so that equivalent measurement represents the mean motional characteristics of group-targets. Under the condition of dense multi-targets, there is no considerable maneuver unless for formation stunt shows.

The group-target tracking in this book is based on evenly variable motional models, so this section will only discuss the classical evenly variable motional model. For the even motion of evenly variable motion and the transformation between maneuver and non-maneuver models, the reader is referred to relevant references.

In a right-angle coordinate system, set the state equation to be $X(k) = \Phi(k, k-1)X(k-1) + G(k-1)W(k-1)$. The evenly variable motional model is a second-moment model. Its 1-dimensional, 2-dimensional, and 3-dimensional state equations are as follows:

(1) 1-dimensional state equation

$$
\begin{bmatrix} x(k) \\ \dot{x}(k) \\ \ddot{x}(k) \end{bmatrix} = \begin{bmatrix} 1 & T & \frac{1}{2}T^2 \\ 0 & 1 & T \\ 0 & 0 & 1 \end{bmatrix} \begin{bmatrix} x(k-1) \\ \dot{x}(k-1) \\ \ddot{x}(k-1) \end{bmatrix} + \begin{bmatrix} \frac{1}{2}T^2 \\ T \\ 1 \end{bmatrix} w_x \qquad (2.31)
$$

(2) 2-dimensional state equation

$$
\begin{bmatrix} x(k) \\ \ddot{x}(k) \\ \dddot{x}(k) \\ y(k) \\ \dot{y}(k) \\ \ddot{y}(k) \end{bmatrix} = \begin{bmatrix} 1 & T & \frac{1}{2}T^2 & 0 & 0 & 0 \\ 0 & 1 & T & 0 & 0 & 0 \\ 0 & 0 & 1 & 0 & 0 & 0 \\ 0 & 0 & 0 & 1 & T & \frac{1}{2}T^2 \\ 0 & 0 & 0 & 0 & 1 & T \\ 0 & 0 & 0 & 0 & 0 & 1 \end{bmatrix} \begin{bmatrix} x(k-1) \\ \ddot{x}(k-1) \\ x(k-1) \\ y(k-1) \\ \dot{y}(k-1) \\ \ddot{y}(k-1) \end{bmatrix} + \begin{bmatrix} \frac{1}{2}T^2 & 0 \\ T & 0 \\ 1 & 0 \\ 0 & \frac{1}{2}T^2 \\ 0 & T \\ 0 & 1 \end{bmatrix} \begin{bmatrix} w_x \\ w_y \end{bmatrix}
$$

$$(2.32)$$

(3) 3-dimensional state equation

$$
\begin{bmatrix} x(k) \\ \dot{x}(k) \\ \ddot{x}(k) \\ y(k) \\ \dot{y}(k) \\ \ddot{y}(k) \\ z(k) \\ \dot{z}(k) \\ \ddot{z}(k) \end{bmatrix} = \begin{bmatrix} 1 & T & \frac{1}{2}T^2 & 0 & 0 & 0 & 0 & 0 & 0 \\ 0 & 1 & T & 0 & 0 & 0 & 0 & 0 & 0 \\ 0 & 0 & 1 & 0 & 0 & 0 & 0 & 0 & 0 \\ 0 & 0 & 0 & 1 & T & \frac{1}{2}T^2 & 0 & 0 & 0 \\ 0 & 0 & 0 & 0 & 1 & T & 0 & 0 & 0 \\ 0 & 0 & 0 & 0 & 0 & 1 & 0 & 0 & 0 \\ 0 & 0 & 0 & 0 & 0 & 0 & 1 & T & \frac{1}{2}T^2 \\ 0 & 0 & 0 & 0 & 0 & 0 & 0 & 1 & T \\ 0 & 0 & 0 & 0 & 0 & 0 & 0 & 0 & 1 \end{bmatrix} \begin{bmatrix} x(k-1) \\ \dot{x}(k-1) \\ \ddot{x}(k-1) \\ y(k-1) \\ \dot{y}(k-1) \\ \ddot{y}(k-1) \\ z(k-1) \\ \dot{z}(k-1) \\ \ddot{z}(k-1) \end{bmatrix} + \begin{bmatrix} \frac{1}{2}T^2 & 0 & 0 \\ T & 0 & 0 \\ 1 & 0 & 0 \\ 0 & \frac{1}{2}T^2 & 0 \\ 0 & T & 0 \\ 0 & 1 & 0 \\ 0 & 0 & \frac{1}{2}T^2 \\ 0 & 0 & T \\ 0 & 0 & 1 \end{bmatrix} \begin{bmatrix} w_x \\ w_y \\ w_z \end{bmatrix}
$$

$$(2.33)$$

where $x(k), \dot{x}(k), \ddot{x}(k), y(k), \dot{y}(k), \ddot{y}(k), z(k), \dot{z}(k)$, and $\ddot{z}(k)$ represent the location, velocity, and acceleration of motional targets in directions of x, y, and z, respectively. T is the sampling cycle. w_x, w_y, w_z are independent Gaussian white noises with zero mean and variance $\sigma_{w_x}, \sigma_{w_y}, \sigma_{w_z}$. The sampling cycle and acceleration variance have considerable influence on the performance of filters. We can choose sampling cycle, but the acceleration variance is an unknown quantity which is difficult to determine.

2.4 Basic Algorithms for Multi-Target Tracking

2.4.1 Review

Data association and track maintenance are the primary problems to be solved in multi-target tracking, currently including maximum likelihood method and Bayesian methods [67, 68, 71]. The maximum likelihood method includes mainly artificial plotting method, track bifurcation method, joint likelihood method, 0–1 integer programming method, and association method. Bayesian methods include Singer's nearest-neighboring method, the all-neighboring optimal method of Singer, Sea, and Housewright, the probabilistic data association method proposed of Shalom, Jaffer, and Tse, Shalom's joint probabilistic data association method, Reid's multiple hypotheses method, Blom's multiple model method, and the interacting multiple model of Blom and Shalom [1, 28, 33, 35, 37]. Recently, L. Hong proposed a interacting multi-velocity multiple model method based on the wavelet transform [46–53].

For the moment, Bayesian algorithms, based on Bayesian rules, still play the leading role and can be categorized into two classes. (1) The optimal Bayesian algorithms process the set of all the validated measurements before current time point, give the probability of each measurement sequence, and synthesize. This class includes the optimal Bayesian algorithm and multiple hypotheses methods and has considerable complexity and computation cost. (2) The second-optimal Bayesian algorithms process only the set of all the validated measurements at current time point. This class includes the nearest-neighboring algorithm, probabilistic data association algorithm, and joint probabilistic data association algorithm and has comparatively low computation cost. The aim of track association (or track identification) is not only to build up an archive of the targets being tracked, but also to complete tasks such as navigation or attack by utilizing target track information. Therefore, although the optimal Bayesian algorithms have many merits compared to the second-optimal Bayesian algorithms, its complexity and high computation cost limit the instantaneity, which is unbearable in realistic applications. In a sense, the association algorithms with good instantaneity and appropriate computation cost have the most powerful vitality [1].

Since group-target tracking is a kind of dense multi-target tracking algorithm based on Bayesian algorithms, next we will emphasize on the nearest-neighboring method, probability data association algorithm, and joint probability data association algorithm.

2.4.2 The Nearest-Neighboring Association Algorithm

In 1971, Singer et al. proposed a tracking filter which had fixed memory demand and could work under condition of multi-returns [37]. This filtering algorithm is the earliest, simplest, and most effective method so far. It takes the statistically "nearest-neighboring" observation data falling inside tracking gates, compared to the predicted location of the targets under tracking, as correlated observation data (although in 1973 Singer and Sea extended the work and developed a class of optimal tracking filters based on the relevant characters of prior statistical estimation [35], which still fell inside the domain of the nearest-neighboring filters). The tracking gates, predicted centers of tracks and the observation data in this cycle, and the relationships among them are shown in Fig. 2.2.

Assume that tracks have been built up before time $k - 1$, and the new measurements at time k are $Y_i(k)$, $i = 1, 2, \ldots, N$. The difference vector between the ith measurement and tracks is defined as the difference between measurements and predicted values. Namely, the filter residual is:

$$v_i(k) = Y_i(k) - H(k)\widehat{X}(k/k - 1) \tag{2.34}$$

Fig. 2.2 Demonstration of the relationships among tracking gates, predicted centers, and measurements

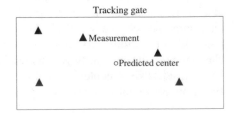

According to Eq. 2.27, $S(k)$ is the covariance matrix of the residual vector. Then, the statistical distance is:

$$
\begin{aligned}
g_i(k) &= \left\| Y_i(k) - H(k)\widehat{X}(k/k-1) \right\|^2 \\
&= \left[Y_i(k) - H(k)\widehat{X}(k/k-1) \right]^T S^{-1}(k) \cdot \left[Y_i(k) - H(k)\widehat{X}(k/k-1) \right]
\end{aligned} \quad (2.35)
$$

$g_0 = \min(g_i)$ is the observation data correlated with tracks, where $i = 1, 2, \ldots, N$.

The nearest-neighboring data association algorithm is mainly applicable to tracking single target or sparse multi-targets. The merits lie in low computation cost and implementation easiness, and the demerit lies in the easy occurrence of association mistakes when returns have comparatively high density [2].

Generally, the applications of the nearest-neighboring filtering algorithm should obey the principles as follows: (1) There is only one measurement within the association gate of some track, and then, the measurement is naturally correlated to the track. (2) Some measurement falls only within one association gate of tracks, and then, the measurement is correlated to the track. (3) There are multiple measurements within the association gate of some track, and then, the measurement which is nearest to the predicted location is correlated to the track. (4) Some measurement falls within multiple association gates of tracks, and then, the measurement is correlated to the nearest track. Reference [2] believes that the solution to these four principles is not sole and is relevant to the successive sequence of measurements. And the solution considers not the situation when there is no correct measurement. In this case, the recent returns that are screened out are unnecessarily true measurements, which can lead to association mistakes.

2.4.3 The Probabilistic Data Association Algorithm

During the process of target tracking, there are real target returns and other returns caused by receiver noises and all kinds of disturbing among the returns that radar detects. Likewise, there are real target observation data and false observation data caused by noise wave residual, receiver noises, and disturbances among the observation data output by radar collectors. These false observation data are all called false measurements (or false observation data). Since true measurements and false measurements are mixed together, we need to figure out which measurement or which set of measurements are derived from real targets when updating filter states. The nearest-neighboring algorithm believes that all returns are effective and the returns nearest to predicted locations are real target returns, and considers not the possibility that all returns are all false measurements, which makes easy occurrence of association mistakes. In 1972, Bar-Shalom and Jaffer proposed a probabilistic data association algorithm for tracking single targets, by utilizing all returns within tracking gates [19]. In this algorithm, all returns within tracking gates are regarded as valid returns; each return can derive from real targets but with different probabilities. This method comprehensively considers all returns within tracking gates, computes the weighted coefficients of each return probability and the weighted sum according to substantial relevant situations, and uses equivalent measurement after weighted summation to update filters.

As we know, data association needs primarily to build up the tracking gates of targets and validates candidate returns based on the tracking gates. According to Kalman filter equations, filtering residual is $v_i = Y(k) - H(k)\widehat{X}(k/k - 1)$, and its covariance is $S(k)$. Assume that real measurements at time k obey Gauss distribution, and then, we get:

$$p[Y(k)/Y^{k-1}] = N[Y(k); \widehat{Y}(k/k - 1), S(k)] \tag{2.36}$$

The tracking gate is defined as:

$$V_k(\gamma) = \left\{ Y(k) : \left[Y_i(k) - H(k)\widehat{X}(k/k - 1)\right]^T S^{-1}(k) \cdot \left[Y_i(k) - H(k)\widehat{X}(k/k - 1)\right] \leq \gamma \right\} \tag{2.37}$$

In Eq. 2.37, γ is a threshold relevant to radar measuring precision and dynamic errors of filters.

The measurements fall inside tracking gates at time k are denoted by:

$$Y(k) = \{Y_i(k)\}_{i=1}^{m_k} \tag{2.38}$$

In Eq. 2.38, m_k is the number of measurements within tracking gates, and it is a random variable. The accumulative measurement is:

$$Y^k = \{Y(j)\}_{j=1}^{k} \tag{2.39}$$

To estimate the target state according to the measurements at the current time point, the measurements before time k can be computed, by assuming that the state vector $X(k)$ at time k based on the condition Y^{k-1} satisfies the second-optimal Bayesian rules.

$$p\left[X(k)/Y^{k-1}\right] = N\left[X(k); \widehat{X}(k/k - 1), P(k/k - 1)\right] \tag{2.40}$$

Thus, we can realize recursive computation. Considering the possibility that no measurement is true, we define:

$$\begin{cases} \theta_i(k) = \{Y_i(k) - \text{Measurement derived from targets}\} \\ \theta_0(k) = \{\text{Measurement at time } k \text{ that derived not from targets}\} \\ i = 1, 2, \ldots, m_k \end{cases} \tag{2.41}$$

The conditional probability under the condition of Y^k is:

$$\beta_i = p\{\theta_i(k)/Y^k\} \quad i = 1, 2, \ldots, m_k \tag{2.42}$$

These events are mutually exclusive and complete. Then, we get:

$$\sum_{i=0}^{m_k} \beta_i(k) = 1 \tag{2.43}$$

According to the all-probability theorem, we get the conditional mean at time k is:

$$\widehat{X}(k/k) = E[X(k)/Y^k] = \sum_{i=0}^{m_k} E[X(k)/\theta_i(k), Y^k] p\{\theta_i(k)/Y^k\} = \sum_{i=0}^{m_k} \beta_i(k) X_i(k/k) \tag{2.44}$$

In Eq. 2.44, $\widehat{X}_i(k/k)$ is the corrected state estimation under condition of event $\theta_i(k)$.

$$\widehat{X}_i(k) = \widehat{X}(k/k-1) + K(k)v_i(k) \quad i = 1, 2, \ldots, m_k \tag{2.45}$$

In Eq. 2.45, $v_i(k)$ and $K(k)$ are the residual and gains output by standard Kalman filter, respectively. Under the condition of $\theta_i(k)$ ($i \neq 0$), measurements are derived from targets and are correct measurements. Considering the possibility that no measurement is correct, we substitute measurements with predicted values, namely the state estimation at this moment, is:

$$\widehat{X}_0(k/k) = \widehat{X}(k/k-1) \tag{2.46}$$

We substitute Eqs. 2.45 and 2.46 into Eq. 2.44 and get the state correction equation of probabilistic data association as follows:

$$\widehat{X}(k/k) = \widehat{X}(k/k-1) + K(k)v(k) \tag{2.47}$$

In Eq. 2.47, $v(k) = \sum_{i=1}^{m_k} \beta_i(k)v_i(k)$. Please note that $v(k)$ is formally linear, but essentially nonlinear, because of its dependence on $\beta_i(k)$. The corresponding error estimation covariance is:

$$P(k/k) - \beta_0(k)\Gamma(k/k-1) + [1 - \beta_0(k)]P^c(k/k) + \widetilde{P}(k) \tag{2.48}$$

In Eq. 2.48,

$$\widetilde{P}(k) = K(k)\left[\sum_{i=1}^{m_k} \beta_i(k)v_i(k)v_i^T - v(k)v^T(k)\right]K^T(k) \tag{2.49}$$

$$P^c(k/k) = [I - K(k)H(k)]p(k/k-1) \tag{2.50}$$

Next, we give the computation method of $\beta_i(k)$ [1, 21, 24–36, 68]. The assumptions made for probabilistic data association algorithm are as follows.

First, tracks are already initiated. Second, there is at most one true measurement among all the measurements at every time point, and the occurrence probability of this even is P_D. Third, correct measurements obey Gauss distribution. Fourth, false measurements all obey uniform distribution.

True measurements are:

$$P[Y_i(k)/\theta_i(k), Y^{k-1}] = f[Y_i(k)/Y^{k-1}] = P_G^{-1}N[v_i(k), 0, S(k)] \quad i = 1, 2, \ldots, m_k \tag{2.51}$$

In Eq. 2.51

$$N[v_i(k), 0, S(k)] = |2\pi S(k)|^{-\frac{1}{2}}\exp\left[-\frac{1}{2}v_i^T(k)S^{-1}(k)v_i(k)\right] \tag{2.52}$$

$$P[Y(k); \theta i(k), m_k, Y^{k-1}] = \begin{cases} V_k^{-m_k}P_G^{-1}|2\pi S(k)|^{-\frac{1}{2}}\exp\left[-\frac{1}{2}v_i^T(k)S^{-1}(k)v_i(k)\right] \\ V_k^{-m_k} \end{cases} \tag{2.53}$$

P_G is the probability that correct measurements fall inside tracking gates. For simplicity, it is set to be 1.

False measurements that fall inside tracking gates are independent from each other and obey uniform distribution.

$$P[Y_i(k); \theta_j(k), Y^{k-1}] = V_k^{-1}, \quad i \neq j \tag{2.54}$$

In Eq. 2.54, V_k is the volume of association gates.

According to Eq. 2.42, we get

$$\beta_i(k) = p\{\theta_i(k)/Y^k\} = P\{\theta_i(k)/Y(k), m_k, Y^{k-1}\} \quad i = 1, 2, \ldots, m_k \tag{2.55}$$

According to Bayesian theorem and multiplication theorem, we can get:

$$\beta_i(k) = \frac{1}{c}p[Y(k)/\theta_i(k), m_k, Y^{k-1}]P\{\theta_i(k)/m_k, Y^{k-1}\} \quad i = 1, 2, \ldots, m_k \tag{2.56}$$

In Eq. 2.56

$$c = \sum_{i=0}^{mk}p[Y(k)/\theta_i(k), m_k, Y^{k-1}]P\{\theta_i(k)/m_k, Y^{k-1}\} \tag{2.57}$$

where c is a normalization constant. Since $i = 0$, all possible events are considered.

When $i = 0$, all measurements are false measurements. According to Hypothesis 4, if the valid measurements before time k and m_k measurements at time k are all derived from noise waves, the probability density of $Y(k)$ is:

$$P[Y(k); \theta_0(k), m_k, Y^{k-1}] = \prod_{i=1}^{m_k} P[Y_i(k); \theta_0(k), m_k, Y^{k-1}] = V_k^{-m_k} \quad (2.58)$$

When $i = 1, 2, \ldots, m_k$, according to Hypothesis 3, if there is certainly one measurement, among the valid measurements before time k and m_k measurements at time k, that is derived from targets, the probability density of $Y_i(k)$ is:

$$p[Y_i(k)/\theta_i(k), m_k, Y^{k-1}] = P_G^{-1} N[Y_i\pi(k); \widehat{Y}(k/k-1), S(k)]$$

$$= P_G^{-1} N[v_i(k); 0, S(k)] = P_G^{-1}(2\pi)^{-\frac{1}{2}} \cdot |S(k)|^{-\frac{1}{2}} \exp\left[-\frac{1}{2} v_i^T(k) S^{-1}(k) v_i(k)\right]$$

$$(2.59)$$

Then, we get the association probability density of $Y(k)$:

$$P[Y(k); \theta_i(k), m_k, Y^{k-1}] = P[Y_i(k); \theta_0(k), Y^{k-1}] \prod_{i=1}^{m_k} P[Y_i(k); \theta_0(k), m_k, Y^{k-1}]$$

$$= V_k^{-m_k} P_G^{-1} |2\pi S(k)|^{-\frac{1}{2}} \exp\left[-\frac{1}{2} v_i^T(k) S^{-1}(k) v_i(k)\right]$$

$$(2.60)$$

Integrating Eqs. 2.55 and 2.57, we can get:

$$P[Y(k); \theta_i(k), m_k, Y^{k-1}] = \begin{cases} V_k^{-m_k} P_G^{-1} |2\pi S(k)|^{-\frac{1}{2}} \exp\left[-\frac{1}{2} v_i^T(k) S^{-1}(k) v_i(k)\right] \\ V_k^{-m_k} \end{cases}$$

$$(2.61)$$

Substituting Eq. 2.58 into Eq. 2.53, we can get:

$$\beta_i(k) = \frac{V_k P_G^{-1} |2\pi S(k)|^{-\frac{1}{2}} r_i(m_k) \exp\left[-\frac{1}{2} v_i^T S^{-1}(k) v_i\right]}{r_0(m_k) + \sum_{i=1}^{mk} V_k P_G^{-1} |2\pi S(k)|^{-\frac{1}{2}} r_i(m_k) \exp\left[-\frac{1}{2} v_i^T S^{-1}(k) v_i\right]}, \quad (2.62)$$

$$i = 1, 2, \ldots, m_k$$

$$\beta_0(k) = \frac{r_0(m_k)}{r_0(m_k) + \sum_{i=1}^{mk} V_k P_G^{-1} |2\pi S(k)|^{-\frac{1}{2}} r_i(m_k) \exp\left[-\frac{1}{2} v_i^T S^{-1}(k) v_i\right]} \quad (2.63)$$

In Eqs. 2.61 and 2.62,

$$\gamma_i(m_k) = P[\theta_i(k)/m_k, Y^{k-1}], \quad i = 1, 2, \ldots, m_k \tag{2.64}$$

It is the prior probability of event $\theta_i(k)$ that is correlated with the number of valid measurements. According to reference [2], we directly get:

$$\gamma_i(m_k) = P\left[\theta_i(k)/m^t = m_k, Y^{k-1}\right] = P[\theta_i(k)/m^t = m_k]$$
$$= \begin{cases} \frac{1}{m_k} P_D P_G \left[P_D P_G + (1 - P_D P_G) \frac{\mu_f(m_k)}{\mu_f(m_k-1)}\right]^{-1}, & i = 1, 2, \ldots, m_k \\ (1 - P_D P_G) \frac{\mu_f(m_k)}{\mu_f(m_k-1)} \left[P_D P_G + (1 - P_D P_G) \frac{\mu_f(m_k)}{\mu_f(m_k-1)}\right]^{-1}, & i = 0 \end{cases}$$
$$\tag{2.65}$$

There are two types of models that can be used for the computation of $\mu_f(m_k)$.

One is parameter model. Assume that m^f obeys Poisson distribution with parameter λV_k, then

$$\mu_f(m_k) = P(m^f = m_k) = e^{-\lambda V_k} \frac{(\lambda V_k)^{m_k}}{m_k} \tag{2.66}$$

In Eq. 2.66, λV_k is the number of false measurements within association gates.

The other is nonparameter model. Assuming that m^f obeys uniform distribution, then

$$\mu_f(m_k) = P(m^f = m_k) = \frac{1}{N}, \quad m_k = 1, 2, \ldots, N - 1 \tag{2.67}$$

Using parameter models, we simplify Eq. 2.60 as

$$P\left[\theta_i(k)/m^t = m_k, Y^{k-1}\right] = P[\theta_i(k)/m^t = m_k]$$
$$= \begin{cases} \frac{P_D P_G}{P_D P_G m_k + (1 - P_D P_G)\lambda V_k}, & i = 1, 2, \ldots, m_k \\ \frac{(1 - P_D P_G)\lambda V}{P_D P_G m_k + (1 - P_D P_G)\lambda V_k}, & i = 0 \end{cases} \tag{2.68}$$

Substituting Eq. 2.68 into Eqs. 2.62 and 2.63, we get association the following matrices.

$$\beta_i(k) = \frac{e_i(k)}{b(k) + \sum_{i=1}^{m_k} e_i(k)} \quad i = 1, 2, \ldots, m_k \tag{2.69}$$

$$\beta_0(k) = \frac{b(k)}{b(k) + \sum_{i=1}^{m_k} e_i(k)} \quad i = 1, 2, \ldots, m_k \tag{2.70}$$

In Eq. 2.70,

$$e_i(k) = \exp\left[-\frac{1}{2}v_i^T S^{-1}(k)v_i(k)\right], \quad i = 1, 2, \ldots, m_k \tag{2.71}$$

$$b(k) = \lambda |2\pi S(k)|^{\frac{1}{2}}(1 - P_D P_G)/P_D, \quad i = 1, 2, \ldots, m_k \tag{2.72}$$

2.4.4   The Joint Probabilistic Data Association Algorithm

When measurements fall within the intersecting areas of different target tracking gates, measurements can be derived from any target, and then, we need to compute the association probability [22, 25, 28] of the association between each measurement and all kinds of possible source targets. In order to get the association probability, we do not set an independent tracking gate for each target and make the consistency between tracking gates and the whole surveillance area. The aim is to make equal the probability density function of each false measurement within the whole surveillance area and thus to get the conditional probability of association events. But this method does not ignore those events that should be ignored and consequently increases extra computation burden. Therefore, Bar-Shalom introduced the conception of cluster matrix. The structure of cluster matrix is:

$$\Omega = [\omega_{jt}] = \begin{bmatrix} \omega_{10} & \cdots & \omega_{1T} \\ \vdots & \ddots & \vdots \\ \omega_{m_k 0} & \cdots & \omega_{m_k T} \end{bmatrix} \tag{2.73}$$

In Eq. 2.73, ω_{jt} is a binary variable. When $\omega_{jt} = 1$, measurement $j(j = 1, 2, \ldots, m_k)$ falls inside the validation gate of target $t(t = 0, 1, \ldots, T)$. Meanwhile, when $\omega_{jt} = 0$, measurement j does not fall inside the validation gate of target t. When $t = 0$, there is no targets, and then the column elements ω_{j0} that Ω is corresponding to is all 1. This is because each measurement can be derived from noise wave or false alarm. Therefore, the purpose of ignoring extra events and decreasing computational burden can be achieved. Since the measurement that falls inside each target tracking gate is regarded as valid measurement, the above-mentioned methods correspondingly obtain the cluster matrix (or called validation matrix) including m_k measurements and T targets. "Cluster" is defined as the maximum set of intersecting tracking gates. Targets are categorized into different groups according to different "clusters" [22]. There is always a matrix with dual-value elements to correlate to each group like this. This matrix is called the cluster matrix.

For any given multi-target problem, once the validation matrix Ω, which reflects the association status between valid returns and targets or noise waves, is given, we can get the association matrix representing all association events by splitting the validation matrix. The two basic assumptions for the splitting are as follows:

Assumption 1 Each measurement has an only source, which means that each measurement is derived from either targets, or noise waves or false alarms. That is to say, we do not consider the situation when one target has multiple returns.

Assumption 2 For any given target, there is only one measurement that is derived from it. If one target is correlated with multiple measurements, only one of the measurements will be selected and set to be true, and others are false. This is consistent with the assumptions made for PDA algorithm.

The basic idea of JPDA algorithm lies in that all valid measurements that fall inside the tracking gates of different targets t can be derived from target t, but the values of association probability of each measurement are different. Thus, we first define association event and then define joint association event.

The association event between the ith measurement and target t is defined as:

$$\theta_{it}(k) = \{\text{Valid measurement } Y_i(k) \text{ is derived from targets}\},$$
$$i = 1, 2, \ldots, m_k; \quad t = 0, 1, 2, \ldots, T \tag{2.74}$$

When t $= 0$, $\theta_{i0}(k)$ denotes the event that measurement $Y_i(k)$ is derived from false measurements. According to the assumptions for PDA algorithm, the event that measurements are associated with target t has the following characters:

(1) Mutual exclusivity:

$$\theta_{it}(k) \bigcap \theta_{jt}(k) = \emptyset, \quad i \neq j \tag{2.75}$$

(2) Completeness:

$$\bigcup_{i=0}^{m_k} P[\theta_{it}(k)/Y^k] = 1 \tag{2.76}$$

The association probability of association event is:

$$\beta_{it}(k) = P[\theta_{it}(k)/Y^k], \quad i = 0, 1, 2, \ldots, m_k; \quad t = 0, 1, 2, \ldots, T \tag{2.77}$$

$\beta_{it}(k)$ denotes the probability value when the ith measurement is associated with target t. According to all-probability equation, we can get:

$$\widehat{X}^t(k/k) = E[X^t(k)/Y^k] = \sum_{i=0}^{m_k} E[X^t(k)/\theta_{it}(k), Y^k]P[\theta_{it}(k)/Y^k] = \sum_{i=0}^{m_k} \beta_{it}(k)\widehat{X}_i^t(k/k)$$
$$\tag{2.78}$$

In Eq. 2.78,

$$\sum_{i=0}^{m_k} X_i(k) = E[X^t(k)/\theta_{it}(k), Y^k], \quad i = 0, 1, 2, \ldots, m_k \tag{2.79}$$

denotes the state estimation of the ith measurement at time k when we perform Kalman filtering on it and target t. $\widehat{X}_0^t(k/k)$ denotes the situation when there is no measurement is derived from targets at time k. We must use the predicted value $\widehat{X}^t(k/k-1)$ as measurement values, namely $\widehat{X}_0^t(k/k) = \widehat{X}^t(k/k-1)$.

Next, we define joint events and the association probability of joint events, according to the definition of association events.

Let $\theta(k) = \{\theta_j(k)\}_{j=1}^{\theta_k}$ be the set of all possible association events at time k, and θ_k is the number of association events in $\theta(k)$. Then,

$$\theta_j(k) = \bigcap_{i=0}^{m_k} \theta_{it_i}^j(k), \quad i = 0, 1, 2, \ldots, m_k; \ j = 0, 1, 2, \ldots, \theta_k; \ t_i = 0, 1, 2, \ldots, T \tag{2.80}$$

$\theta_j(k)$ denotes the jth association event, $\theta_{it}^j(k)$ denotes the event that measurement i is derived from target t_i in the jth association event, and $\theta_{i0}^j(k)$ denotes the event that measurement i is derived from false measurements in the jth association event. Thus, according to Eq. 2.74, we get:

$$\theta_{it}(k) = \bigcup_{j=1}^{\theta_k} \theta_{it_i}^j, \quad i = 0, 1, 2, \ldots, m_k; \ j = 0, 1, 2, \ldots, \theta_k; \ t_i = 0, 1, 2, \ldots, T \tag{2.81}$$

There are two known assumptions for joint probabilistic event data association. First, every measurement has a sole source. Namely, any measurement is derived from either targets, or noise waves or false alarms. In other words, the situation when one target has multiple returns and they are unidentified is not considered. Second, for any given target, at most one measurement is derived from it. If one target generates multiple measurements, only one of the measurements will be selected and set to be true, and others are false. An event is called a feasible event when the two assumptions are satisfied.

A joint event $\theta_j(k)$ is denoted by the association matrix as:

$$\Omega[\theta_j(k)] = \{\omega_{it}^j[\theta_j(k)]\}, \quad i = 0, 1, 2, \ldots, m_k; \ j = 0, 1, 2, \ldots, \theta_k \tag{2.82}$$

In Eq. 2.82,

$$\omega_{it}^j[\theta_j(k)] = \begin{cases} 1, & \theta_{it}^j \subset \theta_j(k) \\ 0, & \text{else} \end{cases} \tag{2.83}$$

denotes in the jth association event, whether measurement j is derived from target t. According to the above-mentioned assumptions, the association matrix satisfies:

$$\sum_{t=0}^{T} \omega_{it}^j[\theta_j(k)] = 1, \quad i = 0, 1, 2, \ldots, m_k \tag{2.84}$$

$$\sum_{i=0}^{m_k} \omega_{it}^j[\theta_j(k)] \leq 1, \quad t = 0, 1, 2, \ldots, T \tag{2.85}$$

For simplicity, we define two binary variables.
One is measurement association indicator:

$$\tau_i[\theta_j(k)] = \sum_{t=0}^{T} \omega_{it}^j[\theta_j(k)] = \begin{cases} 1, & t > 0 \\ 0, & t = 0 \end{cases}, \quad i = 0, 1, 2, \ldots, m_k \tag{2.86}$$

denotes whether measurement i is associated with a real target in joint event $\theta_j(k)$. Let

$$\tau[\theta_j(k)] = \left\{ \tau_1[\theta_j(k)], \tau_2[\theta_j(k)], \ldots, \tau_{m_k}[\theta_j(k)] \right\} \tag{2.87}$$

Then, $\tau[\theta_j(k)]$ reflects whether any measurement is associated with real targets in some joint event.
The other is a target detection indicator.

$$\phi_t[\theta_j(k)] = \sum_{i=0}^{m_k} \omega_{it}^j[\theta_j(k)] \leq 1$$

$$= \begin{cases} 1, & \text{there is measurement } i \text{ that is associated to real targets} \\ 0, & \text{there is no measurement } i \text{ that is associated to real targets} \end{cases},$$

$$t = 0, 1, 2, \ldots, T$$

$$\tag{2.88}$$

denotes whether any measurement is associated with a target t in joint event $\theta_j(k)$, namely whether target t will be detected or not. Similarly, let:

$$\phi[\theta_j(k)] = \{\phi_1[\theta_j(k)], \phi_2[\theta_j(k)], \ldots, \phi_T[\theta_j(k)]\} \tag{2.89}$$

Then, $\phi[\theta_j(k)]$ denotes the event whether any target is detected in joint event $\theta_j(k)$. Let $\varphi[\theta_j(k)]$ denote the number of false measurements in joint event $\theta_j(k)$, then

$$\varphi[\theta_j(k)] = \sum_{i=1}^{m_k} \{1 - \tau_i[\theta_j(k)]\} \tag{2.90}$$

According to the above-mentioned discussion, once we get the validation matrix of the association situations between valid measurements and targets or noise waves, the feasible matrix of all feasible joint events can be obtained by splitting the validation matrix. According to the two assumptions, the splitting of validation matrix should obey the following two principles.

(1) There can exist only one nonzero element in each row of the validation matrix. This is the only nonzero element in this row of the validation matrix, which means each measurement has a sole source. Thus, Assumption 1 is satisfied.
(2) In the feasible matrix, except the first column, there can exist at most one nonzero element in each column, which means at most one measurement is derived from each target. Thus, Assumption 2 is satisfied.

The feasible matrix has a one-to-one correspondence with feasible joint events. Generally, we obtain the feasible matrix by splitting the validation matrix and then determine the feasible joint events. With the increase of target number and the number of valid measurements, the number of feasible matrices quickly expands exponentially. Meanwhile, the bigger the intersecting areas of association gates are and the bigger the number of intersecting areas is, the bigger the number of feasible matrices will be.

The computation of joint event association probability is as follows.

According to Bayesian rules, the conditional probability of the joint events of all measurements at time k is:

$$P[\theta_j(k)/Y^k] = P[\theta_j(k)/Y(k), Y^{k-1}] = \frac{1}{c} P[Y(k)/\theta_j(k), Y^{k-1}] P[\theta_j(k)/Y^{k-1}]$$
$$= \frac{1}{c} P[Y(k)/\theta_j(k), Y^{k-1}] P[\theta_j(k)]$$

$$\tag{2.91}$$

In Eq. 2.91,

$$c = \sum_{j=1}^{\theta_k} P[Y(k)/\theta_j(k), Y^{k-1}] P[\theta_j(k)] \tag{2.92}$$

Similar to the computation given in Eq. 2.60, we can get:

$$P\big[Y(k)/\theta_j(k), Y^{k-1}\big] = \prod_{i=1}^{m_k} P\big[Y_i(k)/\theta_j(k), Y^{k-1}\big] = \prod_{i=1}^{m_k} P\big[Y_i(k)/\theta_{it}^j(k), Y^{k-1}\big]$$

(2.93)

Since false measurements are assumed to obey uniform distribution, true measurements associated with targets are assumed to obey Gauss distribution and association gates are assumed to be corresponding to the whole surveillance area, namely $P_G = 1$, we can get:

$$P\big[Y_i(k)/\theta_{it}^j(k), Y^{k-1}\big] = \begin{cases} N_t[Y_i(k)] = N\Big[Y_i(k); \widehat{Y}_{t_i}(k/k-1), S_{t_i}(k)\Big], & \tau_i[\theta_j(k)] = 1 \\ V^{-1}, & \tau_i[\theta_j(k)] = 0 \end{cases}$$

(2.94)

In Eq. 2.94, $\widehat{Y}_{t_i}(k/k-1)$ is the predicted value of target t_i and $S_{t_i}(k)$ is the covariance of the corresponding new information. Substituting Eq. 2.94 into Eq. 2.93, we can get:

$$P\big[Y(k)/\theta_j(k), Y^{k-1}\big] = V^{-\varphi[\theta_j(k)]} \prod_{i=1}^{m_k} N_{t_i}[Y_i(k)]^{\tau_i[\theta_j(k)]}$$

(2.95)

As we know, if $\theta_j(k)$ is determined, target detection indicator $\phi[\theta_j(k)]$ and the number of false measurements $\varphi[\theta_j(k)]$ are also determined. Thus,

$$P[\theta_j(k)] = P\{\theta_j(k)/\phi[\theta_j(k)], \varphi[\theta_j(k)]\}P\{\phi[\theta_j(k)], \varphi[\theta_j(k)]\}$$

(2.96)

In fact, if the number of false measurements is determined, joint event $\theta_j(k)$ is also solely determined according to $\phi[\theta_j(k)]$. Then, the number of event including $\varphi[\theta_j(k)]$ false measurements should be $C_{m_k}^{\varphi[\theta_j(k)]}$. The rest $m_k - \varphi[\theta_j(k)]$ true measurements can have $\{m_k - \varphi[\theta_j(k)]\}!$ possible joint events. Therefore, we can get:

$$P\{\theta_j(k)/\phi[\theta_j(k)], \varphi[\theta_j(k)]\} = \frac{1}{C_{m_k}^{\varphi[\theta_j(k)]}\{m_k - \varphi[\theta_j(k)]\}!} = \frac{\varphi[\theta_j(k)]!}{m_k!}$$

(2.97)

Again on the account of

$$P\{\phi[\theta_j(k)], \varphi[\theta_j(k)]\} = \mu_f\{\varphi[\theta_j(k)]\} \prod_{t=1}^{T} (P_D^t)^{\phi[\theta_j(k)]}(1 - P_D^t)^{1-\phi[\theta_j(k)]}$$

(2.98)

In Eq. 2.98, P_D^t denotes the detection probability of target t. $\mu_f\{\varphi[\theta_j(k)]\}$ denotes the distribution function of prior probability of the false measurement number.

Substituting Eqs. 2.97 and 2.98 into Eq. 2.96, we can get the prior probability of joint event $\theta_j(k)$ as:

$$P[\theta_j(k)] = \frac{\varphi[\theta_j(k)]!}{m_k!} \mu_f\{\varphi[\theta_j(k)]\} \prod_{t=1}^{T} (P_D^t)^{\phi[\theta_j(k)]}(1 - P_D^t)^{1-\phi[\theta_j(k)]} \qquad (2.99)$$

Likewise, substituting Eqs. 2.93 and 2.99 into Eq. 2.91, we can get the posterior probability of joint event $\theta_j(k)$ as:

$$
\begin{aligned}
&P[\theta_j(k)/Y^k] \\
&= \frac{\varphi[\theta_j(k)]!}{m_k!} \mu_f\{\varphi[\theta_j(k)]\} V^{-\varphi[\theta_j(k)]} \prod_{i=1}^{m_k} N_{t_i}[Y_i(k)]^{\tau_i[\theta_j(k)]} \prod_{t=1}^{T} (P_D^t)^{\phi[\theta_j(k)]}(1 - P_D^t)^{1-\phi[\theta_j(k)]}
\end{aligned}
$$

$$(2.100)$$

According to the probabilistic data association algorithm, the conditional probabilities of the probability distribution function $\mu_f\{\varphi[\theta_j(k)]\}$ with Poisson-distributed parameters and uniform-distributed nonparameters and the corresponding joint events are:

$$\mu_f\{\varphi[\theta_j(k)]\} = e^{-\lambda V} \frac{(\lambda V)^{\varphi[\theta_j(k)]}}{\varphi[\theta_j(k)]!} \qquad (2.101)$$

Substituting Eq. 2.101 into Eq. 2.100, we can get:

$$P[\theta_j(k)/Y^k] = \frac{\lambda^{\varphi[\theta_j(k)]}}{c'} \prod_{i=1}^{m_k} N_{t_i}[Y_i(k)]^{\tau_i[\theta_j(k)]} \prod_{t=1}^{T} (P_D^t)^{\phi[\theta_j(k)]}(1 - P_D^t)^{1-\phi[\theta_j(k)]} \quad (2.102)$$

In Eq. 2.102, c' is a new normalization constant.

$$P[\theta_j(k)/Y^k] = \frac{1}{c} \frac{\varphi[\theta_j(k)]!}{V^{\varphi[\theta_j(k)]}} \prod_{i=1}^{m_k} N_{t_i}[Y_i(k)]^{\tau_i[\theta_j(k)]} \prod_{t=1}^{T} (P_D^t)^{\phi[\theta_j(k)]}(1 - P_D^t)^{1-\phi[\theta_j(k)]}$$

$$(2.103)$$

Thus, we get the probability when the ith measurement is associated with targets as follows:

$$
\begin{aligned}
\beta_{it}(k) &= P[\theta_{it}(k)/Y^k] \\
&= P\left[\bigcup_{j=1}^{\theta_k} \theta_{it_i}^j / Y^k\right] = \sum_{j=1}^{\theta_k} P[\theta_i(k)/Y^k]\omega_{it}^j[\theta_j(k)], \\
i &= 0,1,2,\ldots,m_k; \quad t = 0,1,2,\ldots,T
\end{aligned}
\qquad (2.104)
$$

The association probability when there is no measurement is derived from targets is:

$$\beta_{0t}(k) = 1 - \sum_{i=1}^{m_k} \beta_{it}(k) \qquad (2.105)$$

After obtaining the joint association probability, we can naturally give its filtering equations. Here, we do not give unnecessary details again.

2.4.5   Generalized Probabilistic Data Association Algorithm

Because the computation of the joint probabilistic data association algorithm is comparatively big, researchers propose multiple improved algorithms of JPDA, mainly including the precise nearest-neighboring data association ENNPDA, the coupling probabilistic data association CPDA, the JPDA algorithm with avoidance of track aggregation, joint synthesized probabilistic data association JIPDA, and comprehensive IJPDA. All these improvements make some assumption based on the feasible rules of JPDA algorithm and are simplified algorithms on the cost of precision and cannot essentially solve the computation cost problem. The feasible rule of JPDA is that one measurement can belong to one target and one target can at most possess one measurement.

In "T. Kirubarajan, Bar-Shalom, K.R. Pattipati. "Multiassignment for Tracking a large Number of Overlapping Objects." IEEE Trans. on Aerospace and Electronic Systems. Vol. 37, No. 1, Jan. 2001, pp. 2–20", the viewpoint of the non-one-to-one corresponding between measurements and tracks was also presented. They implement the multi–multi-correspondence between measurements and tracks through the repeated usages of one-to-one distributions. This is also based on the feasible rule of JPDA, and the repeated usages of one-to-one distributions lead to even bigger computation burden.

The generalized probabilistic data association breaks the limitation of the feasible rule of JPDA, under the condition of non-one-to-one correspondence between measurements and targets. It is another feasibility assumption of joint events, namely the two events of targets and measurements constitute a joint event, so that it is an algorithm under condition of repeated usages of both measurements and targets. Compared to JPDA algorithm, the computation load is decreased as the precision is improved.

(1)   The idea of the algorithm

Corresponding to the conception of JPDA joint association event, we define the joint association event in the new algorithm. In order to be different from that in JPDA algorithm, the joint association event is called joint event. With known number of target T and number of measurements m_k, a joint event consists of the

Fig. 2.3 The matrix
consisting of the
cluster-probability statistical
distance between
measurements and targets

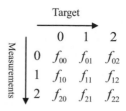

following two events and these two events represent the feasible rule of the new algorithm.

Event ①: Each target has measurements whose number can be several or 0;
Event ②: Each measurement has target sources, whose number can be several or 0.

Here, 0 targets mean no targets. Namely, all measurements are derived from false targets such as disturbances and noise waves, or new targets, except targets under tracking. 0 measurements mean no detection of targets.

Figure 2.3 shows the matrix consisting of the cluster-probability statistical distance between measurements and targets. f_{it} is the statistical distance between measurement i and target t. θ is assumed to be the association event between measurement i and target t.

The events that satisfy events ① and ② are:

$$\theta_{00}\theta_{11}\theta_{22}, \quad \theta_{20}\theta_{21}\theta_{22}, \quad \theta_{00}\theta_{11}\theta_{02}, \quad \cdots$$
$$\theta_{11}\theta_{22}\theta_{00}, \quad \theta_{11}\theta_{21}\theta_{01}, \quad \theta_{12}\theta_{22}\theta_{02}, \quad \theta_{12}\theta_{20}\theta_{02}, \quad \cdots$$

According to the definition, joint events are the combination of the two types of events. We can see that the events that satisfy event ① are on a basis of targets, while the events that satisfy event ② are on a basis of measurements. Thus, a joint event set can be seen as the set of the two types of events. Then utilizing Bayesian rules, we can get the mutual belonging probability β_{it} between measurement i and target t (corresponding to the marginal probability in JPDA algorithm). Although the set of joint events can be partitioned as sets of other types of events, the method discussed before is the simplest, most efficient, and optimal partition.

Figure 2.4 further describes the idea of GPDA algorithm.

The processing on a basis of targets is to deal with the cluster-probability sta-tistical matrix under assumption of "each target has measurements." The problem of "each measurement has its target source" is not considered. In contrary, the pro-cessing on a basis of measurements is to deal with the cluster-probability statistical matrix under the assumption of "each measurement has its target source." Similarly, the problem of "each target has measurements" is not considered.

The difference between GPDA and JPDA lies in whether repeated usage of targets and measurements is allowed. The processing on a basis of targets is

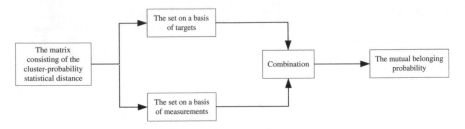

Fig. 2.4 The basic idea of GPDA algorithm

corresponding to the problem of repeated usage of measurements. Namely, "each target has measurements" (these measurements can be the same measurement) solves the problem of "one measurement is associated to multiple targets." Likewise, the processing on a basis of measurements is corresponding to the problem of repeated usage of targets. Namely, it solves the problem of "one target is associated to multiple measurements."

Combining the repeated usage of measurements and targets, we unify "one measurement is corresponding to multiple targets" and "one target is corresponding to multiple measurements" and implement the multi–multi-correspondence between measurements and targets.

(2) Relevant definitions

Like JPDA algorithm, GPDA algorithm is proposed to deal with multi-target tracking problem. The definition of the measurement set is as same as that in JPDA, and so are the set expressions. The validation set at time k is:

$$Y(k) = \{y_i(k)\}_{i=1}^{m_k} \tag{2.106}$$

where m_k is the number of measurements within validation area. Then, the accumulative set of measurements is:

$$Y^k = \{Y(j)\}_{j=1}^{k} \tag{2.107}$$

Similarly, we do not set an independent validation gate for each target and make the consistency between tracking gates and the whole surveillance area. Assume that we track T targets, then m_k is the number of measurements within validation area at time k. We give the following definitions.

Definition 1 $F = [f_{it}]$, $i = 0, 1, \ldots, m_k$, $t = 0, 1, \ldots, T$, where F is the matrix consisting of the cluster-probability statistical distance between measurements and targets. f_{it} is the statistical distance between measurement i and target t, which is also called probability density function.

Definition 2 Under the condition of $F = [f_{it}]$, Θ is the joint events that satisfy new feasibility rules, Θ_t is events that satisfy the assumption ①, and Θ_i is events that satisfy the assumption ②.

$$\Theta = \Theta_t \cup \Theta_i \tag{2.108}$$

(3) Deduction about mutual belonging probability

First construct matrix F, and assume that the state variable of target t obeys normalization distribution with mean of $\widehat{X}_t(k/k-1)$ and covariance of $P_t(k/k-1)$. Namely,

$$P\big[X_t(k)/Y_t^{k-1}\big] = N\Big[X_t(k); \widehat{X}_t(k/k-1), P_t(k/k-1)\Big] \tag{2.109}$$

Then, the probability density function when measurement i $(i \neq 0)$ is corresponding to target t $(t \neq 0)$ (Y. Bar-Shalom, T.E. Fortmann. Tracking and Association. Orlando, FL: Academic Press, 1998) is:

$$\begin{aligned}
f_{it} = p\big[Y_{it}/m_k, Y_t^k\big] &= P_G^{-1} N \pi \Big[Y_{it}(k); \widehat{Y}_t(k/k-1), S_t(k)\Big] \\
&= P_G^{-1} |2\pi S_t(k)|^{-\frac{1}{2}} \exp\Big[-\frac{1}{2} v_{it}^T(k) S_t^{-1}(k) v_{it}(k)\Big]
\end{aligned} \tag{2.110}$$

where $v_{it}(k) = Y_{it}(k) - Y_{it}(k/k\ 1)$.

The probability density function when 0 measurement is corresponding to target t $(t \neq 0)$ means that there is no measurements to be corresponding to the target. Namely, the probability density function of no-detection event is $p_L = f_{0t} = (nV)^{-1}$ $(1 - p_D p_G)$, where n is a proportional coefficient, V is the volume of wave gates, p_D is the detection probability, and p_G is the gate probability.

The probability density function when measurement i $(i \neq 0)$ is corresponding to 0 targets means that this measurement does not belong to any interested target within spatial surveillance area. Namely, the probability density function of the event when this measurement belongs to false targets is:

$$p_F = f_{i0} = \lambda \tag{2.111}$$

where false measurements within wave gates are regarded to obey Poisson distribution and λ is the density of false measurements, namely the number of noise waves within a unit volume.

The association between 0 measurement and 0 target means nothing, and its probability density function is:

$$f_{00} = 0 \tag{2.112}$$

According to the depiction, the matrix $F = [f_{it}]$ consisting of the cluster-probability statistical distance between measurements and targets is:

$$F = [f_{it}] = \begin{bmatrix} & 0 & 1 & 2 & \cdots & T \\ 0 & f_{00} & f_{01} & f_{02} & \cdots & f_{0T} \\ 1 & f_{10} & f_{11} & f_{12} & \cdots & f_{1T} \\ 2 & f_{20} & f_{21} & f_{22} & \cdots & f_{2T} \\ \vdots & \vdots & \vdots & \vdots & \vdots & \vdots \\ m_k & f_{m_k 0} & f_{m_k 1} & f_{m_k 2} & \cdots & f_{m_k T} \end{bmatrix} \qquad (2.113)$$

First, normalizing the matrix given in Eq. 2.113 on a basis of targets, we get the matrix $E_t = [\varepsilon_{it}]$ as:

$$\varepsilon_{it} = \frac{f_{it}}{c_t} \qquad (2.114)$$

where c_t is the normalization constant on target t.

$$c_t = \sum_{i=0}^{m_k} f_{it} \qquad (2.115)$$

Then, normalizing the matrix on a basis of measurements, we get $E_i = [\varepsilon'_{it}]$ as:

$$\varepsilon'_{it} = \frac{f_{it}}{c_i} \qquad (2.116)$$

where c_t is the normalization constant on target t.

$$c_i = \sum_{t=0}^{T} f_{it} \qquad (2.117)$$

The normalized matrix on a basis of targets and the normalized matrix on a basis of measurements are as follows:

$$\begin{bmatrix} & 0 & 1 & 2 & \cdots & T \\ 0 & \varepsilon_{00} & \varepsilon_{01} & \varepsilon_{02} & \cdots & \varepsilon_{0T} \\ 1 & \varepsilon_{10} & \varepsilon_{11} & \varepsilon_{12} & \cdots & \varepsilon_{1T} \\ 2 & \varepsilon_{20} & \varepsilon_{21} & \varepsilon_{22} & \cdots & \varepsilon_{2T} \\ \vdots & \vdots & \vdots & \vdots & \vdots & \vdots \\ m_k & \varepsilon_{m_k 0} & \varepsilon_{m_k 1} & \varepsilon_{m_k 2} & \cdots & \varepsilon_{m_k T} \end{bmatrix} \qquad (2.118)$$

$$\begin{bmatrix} & 0 & 1 & 2 & \cdots & T \\ 0 & \varepsilon'_{00} & \varepsilon'_{01} & \varepsilon'_{02} & \cdots & \varepsilon'_{0T} \\ 1 & \varepsilon'_{10} & \varepsilon'_{11} & \varepsilon'_{12} & \cdots & \varepsilon'_{1T} \\ 2 & \varepsilon'_{20} & \varepsilon'_{21} & \varepsilon'_{22} & \cdots & \varepsilon'_{2T} \\ \vdots & \vdots & \vdots & \vdots & \vdots & \vdots \\ m_k & \varepsilon'_{m_k 0} & \varepsilon'_{m_k 1} & \varepsilon'_{m_k 2} & \cdots & \varepsilon'_{m_k T} \end{bmatrix} \qquad (2.119)$$

Based on E_i, E_t, we give the equation for computing mutual belonging probability.

According to Definition 2, we obviously get:

$$\sum_{f \in F_{it}} p\{\theta_{it}/\Theta, Y^k\} = \sum_{f_1 \in F_{1_{it}}} p\{\theta_{it}/\Theta_t, Y^k\} \cup \sum_{f_2 \in F_{2_{it}}} p\{\theta_{it}/\Theta_i, Y^k\} \qquad (2.120)$$

where the meaning of $\Theta, \Theta_t, \Theta_i$ are as defined in Definition 2, F_{it} is the set of all the joint events f that include θ_{it}, $F_{1_{it}}$ is the set of all the events f_1 that include θ_{it} and satisfy assumption ①, and $F_{2_{it}}$ is the set of all the events f_2 that include θ_{it} and satisfy assumption ②.

According to the conditional Bayesian equation for multiple events, we can get:

$$\begin{aligned} \sum_{f_1 \in F_{1_{it}}} p\{\theta_{it}/\Theta_t, Y^k\} &= \sum_{f_1 \in F_{1_{it}}} p\{\theta_{it}/Y^k\} \cdot p\{\Theta_t/\theta_{it}, Y^k\} \\ &= \sum_{f_1 \in F_{1_{it}}} p\{\theta_{it}/Y(k), Y^{k-1}\} \cdot p\{\Theta_t/\theta_{it}, Y(k), Y^{k-1}\} \\ &= p\left\{\theta_{it}/Y(k), \widehat{X}_t(k/k-1), P_t(k/k-1)\right\} \\ &\quad \cdot \sum_{f_1 \in F_{1_{it}}} p\left\{\Theta_t/\theta_{it}, Y(k), Y^{k-1}, \widehat{X}_t(k/k-1), P_t(k/k-1)\right\} \end{aligned}$$

$$(2.121)$$

Because normalization has been performed before, we pass over the normalization coefficient in this equation. The first item to the right of the last equal mark in Eq. 2.121 is only the prior probability, and its computation equation is:

$$p\left\{\theta_{it}/Y(k), \widehat{X}_t(k/k-1), P_t(k/k-1)\right\} = \varepsilon_{it} \qquad (2.122)$$

64   2 Preliminaries

Because:

$$p\left\{\Theta_t/\theta_{it}, Y(k), Y^{k-1}, \widehat{X}_t(k/k-1), P_t(k/k-1)\right\} = p\left\{\theta_{it}/\Theta_t, Y^k\right\} = \prod_{\substack{r=0,t_r=0 \\ r\neq i, t_r\neq t}}^{m_k,T} \varepsilon_{rt_r}$$

(2.123)

So that,

$$\sum_{f_1\in F_{1_{it}}} p\left\{\Theta_t/\theta_{it}, Y(k), Y^{k-1}, \widehat{X}_t(k/k-1), P_t(k/k-1)\right\} = \sum_{f_1\in F_{1_{it}}} \left(\prod_{\substack{r=0,t_r=0 \\ r\neq i, t_r\neq t}}^{m_k,T} \varepsilon_{rt_r}\right)$$

$$= \prod_{\substack{t_r=0 \\ t_r\neq t}}^{T} \sum_{\substack{r=0 \\ r\neq t}}^{m_k} \varepsilon_{rt_r}$$

Then substituting Eqs. 2.122 and 2.124 into Eq. 2.121, we can get:

$$\sum_{f_1\in F_{1_{it}}} p\left\{\theta_{it}/\Theta_t, Y^k\right\} = \varepsilon_{it} \cdot \prod_{\substack{t_r=0 \\ t_r\neq t}}^{T} \sum_{\substack{r=0 \\ r\neq t}}^{m_k} \varepsilon_{rt_r}$$

(2.125)

where $i = 0, 1, \ldots, m_k$, $t = 0, 1, \ldots, T$.
Likewise we can get:

$$\sum_{f_2\in F_{1_{it}}} p\left\{\theta_{it}/\Theta_t, Y^k\right\} = \varepsilon'_{it} \cdot \prod_{\substack{r=0 \\ r\neq t}}^{m_k} \sum_{\substack{t_r=0 \\ t_r\neq t}}^{T} \varepsilon'_{rt_r}$$

(2.126)

Substituting Eqs. 2.125 and 2.126 into Eq. 2.120, we can get:

$$\sum_{f\in F_{it}} p\left\{\theta_{it}/\Theta_t, Y^k\right\} = \varepsilon_{it} \cdot \prod_{\substack{t_r=0 \\ t_r\neq t}}^{T} \sum_{\substack{r=0 \\ r\neq t}}^{m_k} \cdot\varepsilon_{rt_r} + \varepsilon'_{it} \cdot \prod_{\substack{r=0 \\ r\neq t}}^{m_k} \sum_{\substack{t_r=0 \\ t_r\neq t}}^{T} \cdot\varepsilon'_{rt_r}$$

(2.127)

To guarantee the completeness of the probability of target t, we need to again perform normalization to the same target on the basis of Eq. 2.127 to compute the mutual belonging probability. Then, the mutual belonging probability of target t is:

$$\beta_{it} = \frac{1}{c} \sum_{f \in F_{it}} p\{\theta_{it}/\Theta_t, Y^k\} = \frac{1}{c} \left(\varepsilon_{it} \cdot \prod_{\substack{t_r=0 \\ t_r \neq t}}^{T} \sum_{\substack{r=0 \\ r \neq t}}^{m_k} \cdot \varepsilon_{rt_r} + \varepsilon'_{it} \cdot \prod_{\substack{r=0 \\ r \neq t}}^{m_k} \sum_{\substack{t_r=0 \\ t_r \neq t}}^{T} \cdot \varepsilon'_{rt_r} \right) \quad (2.128)$$

where c is the normalization coefficient for guaranteeing the completeness of measurements within target validation gates.

2.5 Summary

Any new algorithm cannot come out of nothing, so is the group-target tracking algorithm. This chapter first gives the common glossary of multi-target tracking, to provide a reference for the definitions of relevant glossary of multi-target. Then, this chapter gives the deduction procedure about relevant equations of Kalman filtering which is needed by group-target tracking. As for the problem of target maneuver detection and maneuver judgment, based on the depiction on the basic issues of target maneuver detection, this chapter introduces basic mobility models and their three kinds of expressions. Finally, this chapter gives the basic deduction procedures about the nearest-neighboring algorithm, the probabilistic data association algorithm, the joint probabilistic data association algorithm, and the generalized probabilistic data association algorithm, which have important enlightening meaning to group-target tracking, and then introduces the essential method and idea of their implementation. This founds the understanding of group-target and tracking algorithms.

Chapter 3
Grouping Detection and Group Initiation of Group-Target

3.1 Introduction

Multi-targets tracking data association can generate three areas: correctly correlated area, unstable correlated area, and mistaken correlated area [1, 16, 22]. Any radar cannot choose its tracking environment and target distance, but can enhance scan frequency and improve data association algorithm, etc. to decrease unstable correlated area and mistaken correlated area [1, 22]. However small the scan frequency is and however perfect the data association algorithm is, unstable correlated area and mistaken correlated area always exist because tracking gates have to occupy some space. Furthermore, the radar resources limit the arbitrary enhancements of parameters such as scan frequency [2]. Therefore, this book treats identifiable multi-targets in density as group-targets. According to the definition of a group-target, given the restrictions of target distance for group-target formation, targets within surveillance space can always be identified as single target or group-target at any time. That is to say, either we perform single-target tracking on condition that correct association is satisfied, or we perform group-target tracking on condition that group-target formation is satisfied (at least within a certain time period).

The idea of determining the target distance principle of group-targets is as follows. Find some critical distance value under restrictions of radar parameters, to make the predicted values of any two target tracks within surveillance space will not fall inside the tracking gate, the other. Targets satisfying this critical distance value can be identified as single targets (namely sparse multi-targets), otherwise as group-targets.

As an exclusive phase in group-target tracking, grouping detection is an important symbol to distinguish group-target tracking from general multi-target tracking. The basic idea is as follows. First, categorize all the valid measurements in a scan cycle into several sets of observation data, according to target distance principles. Second, take the geometric center of these observation data sets as

© National Defense Industry Press and Springer Science+Business Media Singapore 2017
W. Geng et al., *Group-target Tracking*, DOI 10.1007/978-981-10-1888-6_3

measurement to initiate group-target track. Third, during the course of track initiation, continuously combine the observation data satisfying threshold restrictions and remove those dissatisfying threshold restrictions, to finally get one or more group-targets. As a concurrent procedure to grouping detection, group initiation is not only a process of group-target track initiation, but also a process of rationality validation of grouping detection. The grouping process based on observation data is to categorize all observation data within surveillance space into several sets of observation data, without knowing in advance whether they can form group-targets or not. Therefore, group initiation initiates group-target tracking as validates the rationality of group-target members. Namely grouping detection and group initiation carry out concurrently.

This chapter discusses the determination of group distance principle, grouping detection method, and group initiation algorithm for group-target formation.

3.2   Algorithm of Group-Target Grouping Detection

There are three results from the scan of radar to surveillance space. (1) Only get the observation data of spatial targets when radar opens. (2) Track multi-targets stably and get the tracks of all targets after radar opening. (3) New observation data and stable tracks coexist. During the course of target tracking, we need to investigate the abovementioned three situations concretely. For the first situation, since we get only the number of observation data instead of that of targets, the difficulty is grouping detection with unknown number of targets. For the second and third situations, the number of new observation data, the number of tracks, and their velocities are all known, so grouping detection is comparatively much easier. The essence is to realize the combination of observation data and tracks, tracks and tracks, according to corresponding thresholds. Thus, grouping detection of group-target tracking includes observation data grouping before track initiation and track grouping after track initiation. According to group-target tracking theory, grouping detection of observation data is implemented as an independent process, and this process needs to be combined with group initiation to validate group-target members and initiate tracks. The second and third situations are dealt with as group-target combination and grouping after track initiation. In fact, grouping detection and group-target combination and splitting are two different procedures. The former is performed before track initiation, and the latter is performed after track initiation. For some group-target, once track initiation is completed, grouping detection will stop working (unless new observation data which are uncorrelated to any track appear). When new target appears, these two procedures possibly work concurrently.

This section emphasizes on grouping detection method based on observation data in a more comprehensive sense.

3.2.1   Analysis of Formation Target Grouping Method

As we know, group-target is a new conception based on analyzing formation target conception. The grouping detection method and formation target grouping dealing method are associated but they are also different. Thus, based on the traditional formation target grouping dealing method discussed in foreign and domestic researches, we further analyze the problems existing in formation target grouping, to guarantee the content continuity from formation target grouping dealing method to group-target grouping detection method. And we process similarly in the following research on group initiation, and combination and grouping detection algorithm of group-targets.

Reference [2] discussed the grouping dealing method when there are multiple flying formations with surveillance area. The targets in these formations are in high density, and each formation is within a certain distance of another formation. We call a formation a subgroup (for convenience and consistency, here according to the definition of this book, the group of every formation in other references is called a subgroup; and the remainder of this book will follow this).

Assume the 3-dimensional measurement is Y from one radar scan, and the three components are slope distance R and two sine u and v in two directions. The distance between any two measurements Y_i and Y_j is $d(Y_i, Y_j)$. When measurements Y_i and Y_j are close to each other, we approximately get:

$$d(Y_i, Y_j) = \sqrt{(R_i - R_j)^2 + \bar{R}^2(u_i - u_j)^2 + \bar{R}^2(v_i - v_j)^2} \qquad (3.1)$$

In Eq. 3.1, $\bar{R} = (R_i + R_j)/2$.

Reference [2] firstly gives three definitions.

Definition 1 The distance between any two measurements Y_i and Y_j, which are obtained in the same scan cycle and belong to subgroup U_1 and U_2 respectively, satisfies:

$$d(Y_i, Y_j) < d_0, \quad (d_0 \text{ is a constant}) \qquad (3.2)$$

Then, Y_j belongs to the subgroup to which Y_i belongs.

Definition 2 The distance between any two subgroups U_1 and U_2 within surveillance space is:

$$d(U_1, U_2) = \min_{\substack{Y_i \in U_1 \\ Y_j \in U_2}} d(Y_i, Y_j) \qquad (3.3)$$

According to Eq. 3.2, we obviously know that U_1 and U_2 are two separate subgroups when $d(U_1, U_2) \geq d_0$. Contrarily, they merge into one subgroup.

Definition 3 $Y_1, Y_2, \ldots, Y_n$ are the measurements obtained in the same scan cycle. For any $Y_i (1 \le i \le n)$, let

$$\max_{\substack{1 \le j \le n \\ j \ne i}} d(Y_i, Y_j) < d_0 \qquad (3.4)$$

hold, and $Y_1, Y_2, \ldots, Y_n$ cannot be partitioned as $U_1, U_2, \ldots, U_m (m < n)$. Let the distance between any two subgroups satisfy:

$$d(U_l, U_f) > d_0 \quad (l, f = 1, 2, \ldots, m) \qquad (3.5)$$

Then, $Y_1, Y_2, \ldots, Y_n$ constitute a subgroup. Generally speaking, d_0 is a quantity reflecting the density of targets in a subgroup.

According to the three definitions, the grouping process is as follows.

The first step: check Y_i

If $\underset{i=2,3,\ldots n}{d} (Y_i, Y_j) < d_0$, $Y_i \in Y_j$. Y_i is marked as being checked;

The second step: check Y_k

Select any $Y_k \in Y_j$ and Y_k is unchecked. $\forall Y_i \notin Y_j$, if $d(Y_i, Y_k) < d_0$, $Y_i \in Y_j$, then we mark Y_k as being checked;

The third step: checking terminated

Select any $Y_k \in Y_j$ and Y_k is all checked, then terminate computation. We have got all Y_i that belong to the same subgroup to which Y_j belongs. Otherwise we repeat these steps until all residual measurements have been categorized.

The problems existing in this grouping dealing method is as follows. (1) Step-by-step comparison needs to be performed on the basis of any measurement, namely a serial judgment means is adopted. This will consume a lot of time when there are many targets. (2) The meaning of the short distance between any two measurements is not given, namely the reference for giving target distance values is not given quantitatively.

To solve the abovementioned problems of formation target grouping dealing method such as heavy time consumption for serial splitting target and uncertain target distance definition, etc., we propose grouping detection algorithm based on observation data and group initiation algorithm based on group-target geometric center. The algorithm clarifies the conception of target distance, gives the equation for computing member distance to form group-targets, constructs grouping detection matrix, defines the distance between neighboring group-targets (different from the conception of target distance), and defines the conceptions of cluster-group, group-target, group scale, etc.

3.2.2 Target Distance Principle for Group-Target Formation

Group-target grouping detection faces up to firstly the problem of determining the target distance between members. Only after the establishment of target distance principle on when to track identifiable single targets and when to track group-targets, we can perform grouping detection. Namely the value of member distance of group-targets can be used as a standard to distinguish between identifiable single-target tracking and group-target tracking. The definition must be given unambiguously; otherwise group-target cannot come into being.

Reference [22] defines correctly correlated area when target distance is greater than fivefold standard variance, unstable area when target distance is in the range of twofold to fivefold standard variance, and mistaken correlated area when target distance is less than twofold standard variance. Reference [16] defines sparse multi-targets when target distance is greater than fivefold standard variance, medium dense multi-targets when target distance is threefold standard variance, and dense multi-targets when target distance is equal to standard variance. Namely the three association areas in multi-target tracking are partitioned according to the standard deviation of target distances. Please note here the standard variance represents the variance of the distance between two targets instead of the variance of filter residual, and the mean of random errors of target distances is not equal to zero.

Figure 3.1 displays the partition of multi-target data association areas [1]. The vertical axis shows target distances and the horizontal axis shows scan cycles. Multi-targets tracking data association can generate three areas: correctly correlated area, unstable correlated area, and mistaken correlated area. Within correctly correlated area, multi-targets can be tracked stably, which keeps the good track of each

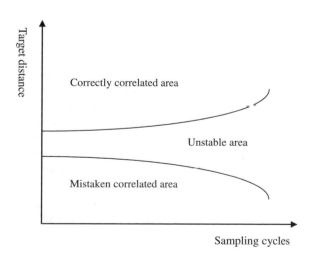

Fig. 3.1 Demonstration of the areas of multi-target data association

target. Within unstable correlated area, multi-targets cannot be tracked precisely; mistaking tracking often happens, system tracking performance degrades, and targets are easy to get lost. Within mistaken correlated area, multi-targets cannot be tracked stably, target tracks drift forward and backward or get mixed (mixed batch phenomenon), but targets are uneasy to get lost. The size of three areas is relevant to not only target distances, but also scan cycles, the size of target detection probability (the size of tracking gate), and data association method. Thus, on precondition of certain target distance and radar measuring precision, increasing correctly correlated area and decreasing unstable correlated area and mistaken correlated area can only be achieved by lessening scan cycles, enhancing detection probability threshold, and improving data association method. However small the scan cycle is (even approaching infinitely small), however big the detection probability threshold is, and however advanced the association algorithm is, there is certain distance between targets and certain measuring precision of radar. Thus, association gates always occupy some space, and unstable correlated area and mistaken correlated area cannot be eliminated completely. This is why group-target tracking is adopted. The aim is to solve the multi-target association ambiguity that cannot be overcome. Group-target tracking is used to implement tracking of multi-target in density, on condition that the tracks of each target cannot be all given.

The physical meaning of group-target tracking can be understood as following. In an adequately long time period, for target tracks that satisfy certain spatial distance (association is difficult within this distance), since they have similar mobility features, we can use the mean mobility state to approximately represent the mobility state of each target. The determination of target distance is relevant to radar measuring precision (including distance, angle, velocity, and acceleration), the prediction precision of filters, data rate of radar (scan cycles), correct reception probability of target returns, and the size of affirmative tracking gate. Therefore, we firstly analyze the concrete meaning and relationships of relevant parameters [16, 22], and then give the definition of target distance of group-targets. For computation simplicity, system errors are assumed to be eliminated in the analysis and the radar measuring precision is the same.

The standard deviations of radar measuring precision [61] mainly include the intrinsic errors of radar itself and target flickering errors. Thus, the errors between observation data only include the standard deviations of radar itself; residual is the difference between measurements and predicted values, its standard deviation includes the standard deviation of radar and the prediction errors of filters. Let σ_o, σ_p represent the standard deviations of radar and filter, respectively. Let $\sigma_d, \sigma_r, \sigma_t$ represent the standard deviations between multiple observation data, observation data and tracks, and tracks and tracks. Radar target capacity is M, and radar is working at full capacity. We get any two measurements $Y_i(k)$ and $Y_j(k)$ at time k, their predicted values are $Y_i(k/k-1)$ and $Y_j(k/k-1)$. The distance between the measurements, the residual between the measurements and their predicted values, and the distance between the predicted values are $d_{ij}^o(k)$, $v_i(k)$, and $v_j(k)$ (the maximum residual vector is $v_{\max}(k)$ among M targets), and $d_{ij}^t(k)$, respectively. They are all represented as the multiples of the standard deviation. Then, we get:

$$d_{ij}^o(k) = K_d \sigma_d = K_d \sqrt{\sigma_{oi} + \sigma_{oj}} = \sqrt{2} K_d \sigma_o \tag{3.6}$$

$$v_i(k) = K_G \sigma_{ri} = K_G \sqrt{\sigma_{oi}^2 + \sigma_{pi}^2} \tag{3.7}$$

$$v_j(k) = K_G \sigma_{rj} = K_G \sqrt{\sigma_{oj}^2 + \sigma_{pj}^2} \tag{3.8}$$

$$d_{ij}^t(k) = K_t \sigma_t = K_t \sqrt{\sigma_{pi}^2 + \sigma_{pj}^2} \tag{3.9}$$

According to Eq. 3.6, the standard deviation mentioned in references [16, 22] is actually σ_d instead of σ_r. Please note that although K_G is the same under condition of equally correct reception probability, the prediction error of each filter is not always equal, so that σ_{ri} is not always equal to σ_{rj}.

If the distance within which the predicted values of two targets do not fall inside the tracking gate of the other is regarded as group-target distance, we cannot guarantee:

$$\begin{cases} d_{ij}^t(k) \geq v_i(k) \\ d_{ij}^t(k) \geq v_j(k) \end{cases} \tag{3.10}$$

can hold at the same time. If multi-group tracking satisfies correct association and stable tracking at the same time, we need to set the maximum residual among M targets to be:

$$v_{\max}(k) = K_G \sigma_{r\max} = \sqrt{\sigma_o^2 + \sigma_{p\max}^2} \tag{3.11}$$

In Eq. 3.11, $\sigma_{p\max} = \max(\sigma_{p1}, \sigma_{p2}, \ldots \sigma_{pM})$. It can guarantee that Eq. 3.10 holds. Even that some targets already have the condition of correct association, we cannot give the distance satisfying all targets. Therefore, without considering the influence of target maneuver errors, set the prediction errors of filters to be equal, namely $\sigma_p = \sigma_{pi} = \sigma_{pj}$, which makes $\sigma_r = \sigma_{ri} = \sigma_{rj}$. When K_t satisfies:

$$K_t \geq \frac{\sigma_r}{\sigma_t} K_G \tag{3.12}$$

We get the classification standard of single target and group-target. The correctly correlated areas of group-target tracking and single-target tracking are shown in Fig. 3.2.

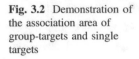

Fig. 3.2 Demonstration of the association area of group-targets and single targets

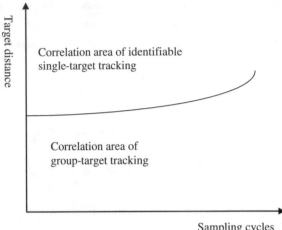

As we know, if adopting the probability data association algorithm to track close multi-targets, any two gargets become disturbance to each other when their tracking gates intersect. This kind of disturbance is small probability event when the intersecting area is comparatively small, and it can only influence measuring precision and tracking stability, but will not be constant disturbance. Deep intersecting appears when the two tracks enter the tracking gate of the other. At this time, each target becomes a constant disturbance to other targets, which makes unstable tracking or mistaken tracking, or even the loss of targets. Therefore, under this circumstance, we usually adopt the association probability data association algorithm which at the same time considers all targets within surveillance area, to implement multi-target tracking. The precondition is that each measurement can be derived from only one target, namely measurements and targets have one-to-one correspondence.

Once target distance becomes shorter, measurements and targets can probably have non-one-to-one correspondence. It is the situation when one measurement can be derived from more than one target that makes the incapability of the association probability data association algorithm to multi-target tracking (namely the precondition of single-target correct tracking is that one track cannot fall inside the tracking gate of others). Considering the close relevance between the size of tracking gate and radar measuring precision, and the Gauss distribution obeyed by real measurements, based on the definition of target density and the categorization of three tracking areas[16, 22], we define the target distance principle to form group-target as follows. **Under condition of same radar precision, full-capacity working of radar, and no maneuver, the critical spatial distance that makes the**

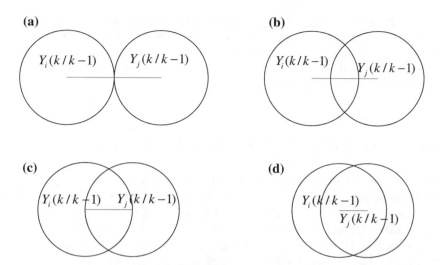

Fig. 3.3 Demonstration of the relationship between target distance and tracking gate. **a** No interaction between correction gates. **b** Local intersecting between correction gates. **c** Critically deep intersecting between correction gates. **d** Deep intersecting between correction gates

predicted values of any two targets not fall inside the tracking gate of the other is called target distance principle. Namely $K_t = \frac{\sigma_r}{\sigma_t} K_G$.

The intuitionistic explanation on the target distance principle to form group-target is as shown in Fig. 3.3 which displays the non-intersecting, local intersecting, critically deep intersecting, and deep intersecting situations, respectively. If target distance in group-target tracking falls within the medium density area, the constant disturbances, which are generated because real measurement and equivalent measurements fall inside some tracking gate, can be eliminated.

Please note that this book calls each member in sparse multi-targets identifiable single target, which is basically consistent with the standard of classifying multi-target density in reference [16, 22]. According to the target distance principle to form group-targets defined in this section, the critical spatial distance that makes the predicted values of any two targets not fall inside the tracking gate of the other is called target distance principle. That is to say, the distance between any two targets should be at least greater than the threshold of tracking gates. The threshold of tracking gates is generally set to be 4–8 time the standard variance of filter residual [1–2]. Obviously, this target distance basically satisfies the categorization standard of sparse multi-targets. We can say that under condition of sparse multi-targets, either tracking gates do not intersect or although they intersect but they are not constant disturbances to each other. Thus, calling each member of sparse multi-targets identifiable single target is practical. In the meantime, this target distance principle for group-targets means that multi-targets can only be categorized as two sets: identifiable single target and unidentifiable multi-target. Thus, during the transition process from dense multi-targets to identifiable single targets, multi-targets in medium density that possibly occur will be merged into group-targets. And identifiable single target and unidentifiable multi-target are processed as sparse multi-targets.

3.2.3 Grouping Detection Based on Observation Data

The basic idea of grouping detection based on observation is as follows. (1) Based on the definition of group-target member distance, take the spatial distance values between every two measurements as elements to construct the grouping detection matrix. (2) In this matrix, an element will be set to be 1 when target distance is less than predefined targets distance $K_t = K_G \sigma_r / \sigma_t$ (defined in the last section), otherwise 0. (3) The observation data corresponding to the elements with value of 1 are regarded as being derived from the same group-target, and some returns corresponding to the elements with value of 0 are called independent observation data and will be processed as false alarms or new targets. The kernel of this grouping detection algorithm is to construct the grouping detection matrix. In order to so, we need to give the following definitions.

Definition 1 According to target distance principles, the set of observation data consisting of the observation data obtained at the moment is called cluster-group.

Definition 2 After the cluster-group forming stable tracks through track initiation, the set of all the observation data within the association gate of this group-target is called group-target.

Definition 3 The distance between the predicted centers of the equivalent measurements of two group-targets is called the distance between group-targets, or abbreviated as group-target distance. Please note that group-target distance is different from the conception of target distance to form group-target. The former is concerning a conception between two group-targets, while the latter is concerning a conception between members of group-target.

Assume that radar receives M valid returns (in measurement coordinate system) from radar scan at one time. Any two measurements are denoted by $Y_i(k) = (R_i, \alpha_i, \beta_i)$ and $Y_j(k) = (R_j, \alpha_j, \beta_j)$ $(i, j = 1, 2, \ldots, M)$. R_i, α_i, β_i and R_j, α_j, β_j are the distance, azimuth, and pitching value of the i, jth measurements, respectively. Then, we get:

$$d_{ij}^t(k) = \left| Y_i(k) - Y_j(k) \right| = \Delta R_{ij}[Y_i(k), Y_j(k)] \tag{3.13}$$

Equation 3.13 gives the spatial distance between measurements.

The grouping detection matrix based on the spatial distance of valid returns is:

$$\begin{bmatrix} & 1 & 2 & \cdots & M \\ 1 & 1 & \ell_{12} & \cdots & \ell_{1M} \\ 2 & \ell_{21} & 1 & \cdots & \ell_{2M} \\ \vdots & \vdots & \vdots & \cdots & \vdots \\ M & \ell_{M1} & \ell_{M2} & \cdots & 1 \end{bmatrix} \tag{3.14}$$

In Eq. 3.14,

$$\ell_{ij} = \begin{cases} 1, d_{ij}^t(k) = \Delta R_{ij}[Y_i(k), Y_j(k)] < K_t \\ 0, d_{ij}^t(k) = \Delta R_{ij}[Y_i(k), Y_j(k)] \geq K_t \end{cases} \quad i,j = 1,2,\ldots,M \tag{3.15}$$

Since $\Delta R_{ij}[Y_i(k), Y_j(k)] = \Delta R_{ji}[Y_j(k), Y_i(k)]$, namely $\ell_{ij} = \ell_{ji}$, the discrimination matrix is upper triangular matrix, and the diagonal elements have value 1 (because $\Delta R_{ii}(k) = 0$ which is definitely less than threshold). According to grouping detection matrix, we get the grouping detection result of measurements within the $L \times L$ 2-dimensional surveillance space. The demonstration is shown as in Fig. 3.4.

According to the Definition 1, the measurements 1, 2, 3, 4, 5, 6, and 7 in Fig. 3.4 constitute subgroup U_1 and measurements 8 and 9 constitute subgroup U_2. This goes in turn. Measurements $M - 1$, $M - 2$, $M - 3$, and $M - 4$ constitute subgroup U_{N-1}, and the independent measurement M constitutes a single group U_N. That is to say, M measurements in total form N subgroups of cluster-group and these subgroups and the single-target forms sparse multi-targets. In other words, we decrease the number of tracks through the grouping detection based on observation data.

We implement grouping based on observation data through the grouping detection matrix and form group-targets. If a restriction of adequately long time (not less than the time of track initiation) is added, the definition of group-target can be also explained mathematically.

	1	2	3	4	5	...	L-4	L-3	L-2	L-1	L
1						...				M-3	
2			8			...			M-1	M-2	
3			9			...			M-4		
4						. .					
5						...					
⋮	⋮	⋮	⋮	⋮	⋮	⋮	⋮	⋮	⋮	⋮	⋮
L-4						...					
L-3		4	3			...					
L-2			1,2	6		...					
L-1		7	5			...					
L						...					M

Fig. 3.4 Demonstration of the spatial distribution of measurements

3.3 Group Initiation Based on Geometric Centers of Group-Targets

Track initiation is an important constituent of multi-target tracking theory [1], and also a necessary decision-making procedure for constructing new target archives. Treating unidentifiable dense multi-targets as group-targets to implement tracking also needs track initiation. Since group-target equivalent measurement (geometric center, centroid or center of gravity) in group-target tracking is regarded as measurements, the precondition of group-target track initiation is to firstly obtain equivalent measurement of group-targets, and the obtaining of equivalent measurement needs to be easy. For elaboration convenience, group-target track initiation is abbreviated as group initiation. Group initiation can adopt two methods. (1) One is to perform track initiation after forming group-targets. The merit is reliable, but it needs a specific validation process of group-targets. (2) The other is to perform track initiation during the course of transition from cluster-groups to group-targets. The merit lies in the short initiation and fast velocity, but it is not very reliable. Radar capture time is a very important parameter; here, we adopt the latter for the sake of saving validation time.

The basic idea of group initiation is as follows. (1) Compute the geometric center after the formation of cluster-groups. (2) Taking the maximum velocity value as the group initiation threshold, search in shape of plum-blossom around the predicted geometric center obtained in the last cycle. (3) Get the cluster track by drawing a line between the geometric center obtained in the last cycle and the geometric center obtained in this cycle, and from the stable track according to the logic track initiation principle. Finally, we implement group initiation as validating the intention of steady group (namely group-target).

Equivalent measurement can represent the holistic mobility law of group-targets. It can be the geometric center, centroid or center of gravity of group-target. Since group-target tracking takes equivalent measurement as its measurements, the equivalent measurement of cluster-group also needs to be obtained during the course of group initiation and the obtaining of equivalent measurement needs to be easy. Therefore, after comprehensively comparing the three possible equivalent measurement of cluster-group, we select the geometric center of cluster-group as equivalent measurement to perform group initiation, in terms of the difficulty degree of obtaining equivalent measurement.

3.3.1 Analysis of Formation Target Track Initiation

The track initiation of formation target performs track initialization after the splitting processing of formation targets. This initialization, on a basis of formation target, assumes the consistency of velocity and direction for members in the same subgroup, adopts the same filter to make each track filtering in the same subgroup have the same gain. The requirement on the basically same velocity and direction for members in the same subgroup is limited only to the initialization phase of tracks and is not asked any more after stable tracking. The track initialization method, given in reference [2], needs to firstly judge the association situation between subgroups in neighboring cycles.

The measurement obtained at time t_1 is divided into m_1 subgroups, denoted by $U_1(t_1), U_2(t_1), \ldots, U_{m_1}(t_1)$. The measurement obtained at time t_2, namely the next cycle, is divided into m_2 subgroups, denoted by $U_1(t_2), U_2(t_2), \ldots, U_{m_2}(t_2)$. In order to judge which two are correlated to each other between the m_1 subgroups obtained at time t_1 and the m_2 subgroups obtained at time t_2, we need to compute the spatial distances between every two subgroup, if the following condition is satisfied:

$$d\big[U_i(t_1), U_j(t_2)\big] < v_{\max}(t_2 - t_1) \quad (i = 1, 2, \ldots, m_1; j = 1, 2, \ldots, m_2) \tag{3.16}$$

In Eq. 3.16,

$$d\big[U_i(t_1), U_j(t_2)\big] = \min d\big[U_i(t_1), U_j(t_2)\big] \tag{3.17}$$

Then, $U_i(t_1)$ and $U_j(t_2)$ are correlated. In Eq. 3.16, $v_{\max}$ is the maximum velocity of targets.

After the association judgment between subgroups in neighboring cycles is finished, we start to process the track initialization of subgroups. There are two situations as follows.

First, we adopt similarity method when the number of subgroups is greater than 3. Thus, formation target has only location displacement instead of distortion in neighboring cycles (with certain allowable error ranges). The basic idea of this method is to find two measurements in $U_i(t_1)$ and draw a line between them, and then find the corresponding measurement line in $U_j(t_2)$. If the two lines are basically in parallel and approximately equal in length, we can get the approximation of the flying velocity v of the subgroup according to the two pairs of measurements.

$$\hat{v} = \big[\overline{Y}(t_2) - \overline{Y}(t_1)/(\bar{t}_2 - \bar{t}_1)\big] \tag{3.18}$$

In Eq. 3.18, $\overline{Y}(t_1) = Y_1(t_1) + Y_2(t_1)/2$ and $\overline{Y}(t_2) = Y_{j1}(t_2) + Y_{j2}(t_2)/2$. Till now, if Eq. 3.18 is the only expression, subgroup track initiation finishes. If Eq. 3.18 is not the only expression, we firstly merge the measurement pairs with approximately equal velocity. When there is only one velocity left, this is the velocity of the subgroup, then track initiation finishes. If there are still many velocity values left after merging, we need to wait for another cycle and take the velocity with the most associations as the approximation of subgroup velocity.

The second track initiation method is called bifurcation method, used for the track initialization when the number of targets in subgroups is equal to or less than 2. The basic idea is similar to that of similarity method, by judging whether Eq. 3.18 is the only expression or not. But this method needs three scan cycles to finish track initiation. Here, we do not give unnecessary details again.

The problems existing in the abovementioned similarity method and bifurcation method to initialize tracks are as follows. (1) We need to perform grouping detection in every cycle during the course of subgroup track initialization. This is effective without considering disturbance and false alarms. But if there are disturbance and false alarms, subgroup initiation will become very difficult. Namely the capability of fault tolerance is not strong. (2) The demand on parallels between measurements of the same subgroup in neighboring cycles is too strict, which leads to low efficiency in track initiation. That is to say, similarity method and bifurcation method essentially are an initiation method for a single measurement, and is applicable to the initiation of strict formation targets but not to the initiation of group-targets defined in this book.

Since the abovementioned methods are not completely applicable to the group initiation of group-targets, we need to find a group initiation algorithm with high fault-tolerance capability, high initiation efficiency, and simple implementation.

3.3.2 Group Initiation Based on Geometric Centers

1. Computation of group-target geometric centers

Group-targets are sets of multi-targets with comparatively fixed spatial locations and basically identical mobility state. Therefore, before computing group-target geometric centers, according to the target distance principle, we need firstly determine which targets are single targets and which possibly form group-targets and how many group-targets they can form, among the multi-targets in density within surveillance space. After finishing group-target filtering and splitting, we can compute concurrently the geometric center of each group-target. It is actually the splitting that transforms dense multi-targets into sparse multi-targets.

Geometric centers of group-targets are the geometric centers within the space enclosed by group-target member returns. For radar with 3 coordinates, we compute the difference between the maximum and the minimum of the return locations in three directions of distance, angle, and pitching within the space, and then construct a cuboid. The half of the difference represents the geometric center of the group-target.

Assume that radar receives M valid returns (in measurement coordinate system) from radar scan at one time. Then, we get:

$$Y_i(k) = (R_i, \alpha_i, \beta_i) \quad (i = 1, 2, \ldots, M) \tag{3.19}$$

In Eq. 3.19, R_i, α_i, β_i are the distance, angle, and pitching of the ith return, respectively.

Let

$$\begin{cases} A_{\max}(k) = (\alpha_1, \alpha_2, \ldots \alpha_M)_{\max} \\ E_{\max}(k) = (\beta_1, \beta_2, \ldots \beta_M)_{\max} \end{cases} \tag{3.20}$$

$$\begin{cases} A_{\min}(k) = (\alpha_1, \alpha_2, \ldots \alpha_M)_{\min} \\ E_{\min}(k) = (\beta_1, \beta_2, \ldots \beta_M)_{\min} \end{cases} \tag{3.21}$$

$$\begin{cases} R_{\max}(k) = (R_{\alpha_{\max}}, R_{\alpha_{\min}}, R_{\beta_{\max}}, R_{\beta_{\min}})_{\max} \\ R_{\min}(k) = (R_{\alpha_{\max}}, R_{\alpha_{\min}}, R_{\beta_{\max}}, R_{\beta_{\min}})_{\min} \end{cases} \tag{3.22}$$

Then, we get:

$$R_{\max}(k) - R_{\min}(k) = \Delta R(k) \tag{3.23}$$

$$A_{\max}(k) - A_{\min}(k) = \Delta A(k) \tag{3.24}$$

$$E_{\max}(k) - E_{\min}(k) = \Delta E(k) \tag{3.25}$$

In Eqs. 3.23, 3.24, and 3.25, $R_{\max}(k)$, $A_{\max}(k)$, $E_{\max}(k)$ represent the maximum of the distance, angle, and pitching under angle restriction, respectively. $R_{\min}(k)$, $A_{\min}(k)$, $E_{\min}(k)$ represent the minimum of the distance, angle and pitching under angle restriction, respectively. Then, the central observation data coordinates are as follows:

$$Y(k) = \left(R_{\min}(k) + \frac{1}{2}\Delta R(k), A_{\min}(k) + \frac{1}{2}\Delta A(k), E_{\min}(k) + \frac{1}{2}\Delta E(k) \right) \tag{3.26}$$

Equation 3.26 gives the geometric center of group-targets.

2. Formation of group initiation threshold

Group initiation threshold is set, taking the maximum velocity of groups to initialize groups. To set its value, we need to consider the scale of group-target members besides the velocity of group-targets. This is different from the setting of the track initiation threshold in multi-target tracking. Assume the maximum velocity of group-target is $V_{\max}$, and radar scan cycle is T.

According to Eqs. 3.23–3.25, the space enclosed by group-targets is a cuboid with the following dimension.

$$G(k) = \Delta D_R(k) \times \Delta D_A(k) \times \Delta D_E(k) \tag{3.27}$$

In Eq. 3.27,

$$\Delta D_R(k) = \Delta R(k) \tag{3.28}$$

$$\Delta D_A(k) = (R_{\alpha\max}(k) + R_{\alpha\min}(k)) \sin\frac{1}{2}\Delta A(k) \tag{3.29}$$

$$\Delta D_E(k) = (R_{\beta\max}(k) + R_{\beta\min}(k)) \sin\frac{1}{2}\Delta E(k) \tag{3.30}$$

Since the maximum velocity of group-target and radar scan cycle is set in advance, according to Eq. 3.27, the group initiation threshold of group-target is:

$$X_{G_n}(k) = \frac{1}{8}\Delta D_R(k) \times \Delta D_A(k) \times \Delta D_E(k) + V_{\max}T \tag{3.31}$$

3. Track initialization of group-targets

We cannot predict the moving direction of group-targets during the group initiation phase. So we take the geometric center of cluster-group as the origin and use five beams to scan in shape of plus-blossom, in order to quickly search the location of cluster-group in the next cycle.

According to Eq. 3.26, the measurements of the geometric centers of cluster-groups at time k and $k+1$ $(k = 1, 2, \ldots, K)$ are as follows:

$$Y(k) = \left(R_{\min}(k) + \frac{1}{2}\Delta R(k), A_{\min}(k) + \frac{1}{2}\Delta A(k), E_{\min}(k) + \frac{1}{2}\Delta E(k) \right) \tag{3.32}$$

$$Y(k+1) = \left(R_{\min}(k+1) + \frac{1}{2}\Delta R(k+1), A_{\min}(k+1) + \frac{1}{2}\Delta A(k+1), E_{\min}(k+1) + \frac{1}{2}\Delta E(k+1) \right) \tag{3.33}$$

We perform initialization association judgment after obtaining the abovementioned two measurements.

$$|Y(k+1) - Y(k)| < V_{\max}T \tag{3.34}$$

If Eq. 3.34 is satisfied, the association is successful, and we can get the rough approximation of velocity.

$$\hat{v} = Y(k+1) - Y(k)/T \tag{3.35}$$

If equivalent measurement is the only association to $Y(k) \leftrightarrows Y(k+1)$, the track initiation has been finished. Next we need to carry on the validation of group-targets and precise approximation of velocity according to predefined principles. If the association is not the only, we need to perform the association judgment in the next cycle, until K association judgments are finished. If it is still not the only, equivalent measurement will be regarded as valid track after track initiation validation and be processed as multi-tracks. Otherwise, we need to redo grouping detection.

3.4   Summary

According to the categorization standard of sparse multi-targets, medium density multi-targets, and dense multi-targets, we define the target distance principle to form group-targets. Under condition of same radar precision, full-capacity working of radar, and no maneuver, the critical spatial distance that makes the predicted values of any two targets not fall inside the tracking gate of the other is called target distance principle. This process can be regarded as the preprocess of measurements, and it is also one of the foundations for studying group-target tracking.

On the basis of the group-target distance definition, we define cluster-group and group-target and perform the grouping detection algorithm research based on observation data. This method takes the spatial distance values between every two measurements as elements to construct the grouping detection matrix and implement the formation of group-targets. We propose the group initiation method based on group-target geometric centers. First, compute the geometric center of the space enclosed by cluster-groups. Second, according to the maximum velocity of group-target and radar scan cycle that are set in advance, we determine the group initiation threshold. Finally, according to the predefined principle, take the geometric center of cluster-groups as measurements to perform group initiation association, until the accomplishment of group initiation and the formation of group-targets. The group initiation algorithm has the function of not only initiating group-target tracks, and also validating the correctness of the transition from cluster-group to group target during the concurrent course of grouping detection. The simulation results prove the correctness and validity of the algorithm, as shown in Figs. 7.7–7.9.

This chapter also clarifies the two basic conceptions of the grouping based on observation data and that based on tracks and indicates the application domains of grouping detection and group-target combination and splitting detection. The grouping detection is applicable to the processing of grouping based on observation data, while the group-target combination and splitting detection is applicable to the processing of grouping detection based on observation data and tracks, and track–track splitting detection.

Chapter 4
Single-Group-Target Data Association and Track Maintenance

4.1 Introduction

As we know, the nearest-neighboring algorithm and probability association algorithm in Bayesian algorithm are very successful two data association algorithms. The nearest-neighboring method, as one of the comparatively effective data association methods, regards the statistically "nearest" observation data that fall inside association gates as the correlated observation data of the target under tracking. The probability data association algorithm is an all-neighboring method. It comprehensively considers all the measurements inside tracking gate, computes the equivalent measurement according to the probabilistic weighted coefficients of each measurement, and updates target state using equivalent measurement. Its key is to get the posterior probability of each candidate measurement deriving from targets under tracking. An important hypothesis of the association algorithm is that there is at most one measurement within tracking gate is true, and the probability density function of real measurements is assumed to obey Gauss distribution (the basis of second-optimal Bayesian estimation) and that of false measurements is assumed to obey uniform distribution. This enables the computation of the posterior probability of each measurement within tracking gates.

As for group-targets, we cannot know precisely the numbers of real targets and false targets of group-targets, and thus cannot know the numbers of real targets and false targets which are derived from targets. Namely we cannot distinguish what measurements are derived from real measurements, and what are derived from false measurements. When real measurements of multi-targets and false measurements of false targets are indistinguishably mixed together, we can only equally process the measurements with diverse probability distributions within association gates of group-targets. PDA algorithm, as an all-neighboring method, assumes that there is at most one real measurement within tracking gates and one target can be only corresponding to one measurement, which is not suitable for the definition of group-targets and real situation. The existence of at most one real measurement within tracking gates is not in accord with the definition of group-target, and targets and measurements are not one-to-one corresponding due to factors such as electromagnetic wave coupling and sheltering between group-target members. Thus, direct adoption of PDA or NNDA algorithm for data association is unsuitable for group-targets. What kind of expression can reflect the two different distributions, describe the contribution of both real measurements and false measurements, and make the two types of measurement compatible with each other? The nearest-neighboring principle tells that the measurement nearest to the predicted center is regarded as real measurements derived from targets. Probability data association principle tells that every measurement within tracking gates is possibly derived from targets, but the probability density of each return is different. The idea of probability data association is consistence with the idea of this book that real measurements can be categorized into direct and indirect measurements. In other words, the non-one-to-one correspondence between multiple real measurement and targets of group-targets is explainable both theoretically and practically. Therefore, according to the definition and features of group-targets, we combine the two ideas of nearest neighboring and all neighboring and propose a nearest-neighboring and all-neighboring association algorithm.

This chapter first defines the conception of double thresholds of association gates and tracking gates of group-targets, describes the method of constructing association gates and tracking gates of group-targets, and then constructs the nearest-neighboring and all-neighboring association algorithm for single-group-targets.

4.2 Association Gates and Tracking Gates of Group-Targets

The formation of tracking gates is the first important problem in multi-target tracking [1]. Tracking gates take the predicted value of radar in the last cycle as center and occupy some spatial area. The dimension of tracking gates decides the probability of receiving correct returns. Tracking gate principle is a rough detection method of pairing measurements and existing tracks or new targets. In many references [62–84,100–125], association gates and tracking gates are common. However, compared to multi-target tracking, group-target takes equivalent measurement as its measurement and needs to separate between the validation of group-target members and equivalent measurement tracking. Namely association gates and tracking gates complete the validation and tracking of group-target members together. Thus, association gates and tracking gates are different in function, conception, and formation method. **The surveillance space volume occupied by all members of group-targets is called the scale of the group**, and its size is up to the contour shaped by group-target marginal members, target distances, the number members. The contour shape and target distance between members directly influence the group scale, and the number of members indirectly influences the group scale. In fact, the contours are different due to the viewing angles of radar beams toward group-targets, together with the combination and splitting of group-targets, so that association gates of group-targets are a spatial area changing with group scale. According to the definition and the information dealing process of group-target tracking algorithm, we give the following hypotheses.

Hypothesis 1: Group-target tracks have been initiated after splitting detection and group initiation.
Hypothesis 2: The returns that fall inside association gates are valid returns and at least two of them are true (consistent with the definition of group-target).
Hypothesis 3: One target can have multiple measurements, and one measurement can be corresponding to multiple targets. Namely targets and measurement have multi–multi correspondence.

Therefore, based on these hypotheses, we give the definitions of the association gates and tracking gates of group-targets.

Definition 1 The spatial area is called a association gate of group-target, if its center is located at the predicted center of equivalent center of group-targets and within it all measurements satisfying target distance principle are covered.

Definition 2 The subspace within the association gate of group-targets is called group-target tracking gates, if its center is located at the predicted center of equivalent center of group-targets and its dimension is decided by the probability of correctly receiving equivalent measurement.

4.2.1   Group-Target Tracking Gates

(1)   Analysis of multi-target tracking gates

In order to form the tracking gates of group-targets, we first review the formation method of tracking gates of traditional multi-target tracking. During the course of multi-target tracking, the main function of tracking gates is to validate candidate returns. A measurement $Y_j(k)$ is obtained at time k, the predicted value of it is $Y_j(k/k-1)$, and the residual vector is $v_j(k)$. The ith component is $y_{ij}(k)$, $y_{ij}(k/k-1)$ and $v_{ij}(k)$, and all the components possess the same tracking gate coefficient K_G. When the residual of measurement $Y_j(k)$ and predicted value $Y_j(k/k-1)$ satisfies:

$$\left| v_{ij}(k) \right| = \left| y_{ij}(k) - y_{ij}(k/k-1) \right| \le K_G \sigma_{ri} \qquad (4.1)$$

$Y_j(k)$ is called a candidate return. In Eq. 3.41, $\sigma_{ri} = \sqrt{\sigma_{oi}^2 + \sigma_{pi}^2}$, where σ_{oi}^2 is the noise variance of the ith component, σ_{pi}^2 is the ith diagonal element of the predicted covariance matrix $P(k/k-1)$, and the value of K_G is decided by the probability of correctly receiving returns. Assuming errors satisfy Gauss distribution and I components and errors are independent from each other (I is the dimension of measurement components), the correct probability [1] that returns fall inside tracking gates is:

$$
\begin{aligned}
P_G &= \int_{V_G} \cdots \int f[v_j(k)]\mathrm{d}v_{ij}\cdots \mathrm{d}v_{Ij} \\
&= \int_{-K_G}^{K_G} \frac{e^{-\frac{u_1}{2}}}{\sqrt{2\pi}}\mathrm{d}u_1 \cdots \int_{-K_G}^{K_G} \frac{e^{-\frac{u_I}{2}}}{\sqrt{2\pi}}\mathrm{d}u_I = [P_r(|t|) \le K_G]^I = [1 - P_r(|t|) > K_G]^I
\end{aligned}
$$

$$(4.2)$$

In Eq. 4.2, V_G is the volume of tracking gates; $P_r[(|t|) \le K_G]$ and $P_r[(|t|) > K_G]$ are the probabilities of receiving and rejecting correct returns, respectively; $u_i = v_{ij}/\sigma_{ri}$; random variable t obeys standard normal distribution $N(0, 1)$. And the volume of I-dimensional tracking gate is:

$$
V_G(I) = \int_{-K_G\sigma_r}^{K_G\sigma_r} \mathrm{d}v_{1j} \cdots \int_{-K_G\sigma_r}^{K_G\sigma_r} \mathrm{d}v_{Ij} = (2K_G)^I \prod_{i=1}^{I} \sigma_{ri}
$$

$$(4.3)$$

(2) Formation of group-target tracking gates

Group-target tracking takes equivalent measurement as its measurement. We first discuss the variance of equivalent measurement.

Since we do not know the concrete number of real measurements in one-time radar scan, we assume that all the measurements within association gates of group-targets include at least two real measurements obeying Gauss distribution with zero mean.

Assume we get M real measurements:

$$
\begin{aligned}
Y^e(k) &= \sum_{j=1}^{M} \beta_j(k)Y_j(k) = \sum_{j=1}^{M} \beta_j(k)\left[Y_{0j}(k) + \sigma_{oj}\right] \\
&= \sum_{j=1}^{M} \beta_j(k)Y_{0j}(k) + \sum_{j=1}^{M} \beta_j(k)\sigma_{oj}
\end{aligned}
$$

$$(4.4)$$

In Eq. 4.4, $\beta_j(k)$ is the weight of each measurement (will be discussed exclusively later) and has completeness. $Y_{0j}(k)$ is the true value of measurement. As for the same phased array radar, its random measurement error is approximately equal, namely $\sigma_{oj} = \sigma_o$. Thus, the last equation turns to be:

$$Y^e(k) = \sum_{j=1}^{M} \beta_j(k) Y_{0j}(k) + \sum_{j=1}^{M} \beta_j(k) \sigma_{oj}$$

$$= \sum_{j=1}^{M} \beta_j(k) Y_{0j}(k) + \sigma_o \sum_{j=1}^{M} \beta_j(k) = \sum_{j=1}^{M} \beta_j(k) Y_{0j}(k) + \sigma_o \qquad (4.5)$$

That is to say, under condition of equal precision, the measurement error of equivalent measurement is changeless, namely $\sigma_o^e = \sigma_o$.

In fact, the M measurements are not all real measurements. Assuming they include only m real measurements, set the measurement equation of false measurement to be $Y_i(k) = H(k)\widehat{X}(k/k - 1) + w_i(k)$, where $\widehat{X}(k/k - 1)$ is the one-step prediction of state vector, and $w_i(k)$ is the white noise obeying uniform distribution inside tracking gates with $H(k)\widehat{X}(k/k - 1)$ as the predicted center. Equation 4.4 turns to be:

$$Y^e(k) = \left[\sum_{j=1}^{m} \beta_j(k) \, Y_{0j}(k) + \sum_{j=1}^{m} \beta_j(k) \, \sigma_o \right] + \left[\sum_{i=1}^{M-m} \beta_i(k) \, Y_i(k) + \sum_{i=1}^{M-m} \beta_i(k) \, w_i \right]$$

$$(4.6)$$

On condition that the number of measurements within association gates, M, is a certain value, $m = 2$ is the case with least real targets instead of most real targets, and thus the variance of equivalent measurement is maximum. Constructing tracking gates with maximum variance can obviously guarantee the correct reception of equivalent measurements. The algorithm given in references [1, 68] computes the equivalent noise of single-target probability data association and can get the variance of equivalent measurement when $m = 2$ as follows.

$$\sigma_o^{e2} = \sigma_o^2(\beta_1 + \beta_2) + \left[(1 - \beta_1)^2 + (1 - \beta_2)^2 \right] H(k) P(k/k - 1) H(k)^T$$

$$+ \sum_{i=3}^{M} \beta_i^2 v_{w_i}^?$$

$$(4.7)$$

Thus, we get:

$$\sigma_{ri}^e = \sqrt{\sigma_{oi}^{e2} + \sigma_{pi}^2} \qquad (4.8)$$

On condition that the I components of equivalent measurement (equivalent measurement got from weighting the multiple measurements of the same radar naturally has I components) and errors are all independent from each other, the tracking which takes group-target equivalent measurement as measurement essentially is single-target tracking. Therefore, the equation for tracking gate volume computation is also suitable for group-target tracking gate computation. With the same tracking gate coefficient and equal all-dimensional variance, the volume of tracking gate of equivalent measurement is:

$$V_{\text{GT}}(I) = \int_{-K_{\text{G}}\sigma_{r1}^e}^{K_{\text{G}}\sigma_{r1}^e} \mathrm{d}v_1^e \cdots \int_{-K_{\text{G}}\sigma_{rI}^e}^{K_{\text{G}}\sigma_{rI}^e} \mathrm{d}v_I^e = (2K_{\text{G}}\sigma_r^e)^I \tag{4.9}$$

where $v_i^e(k) = Y^e(k) - Y^e(k/k-1)$, $i = 1, 2, \ldots, I$. Namely the tracking gate is:

$$\gamma_{\text{T}}(k) = K_{\text{G}}\sigma_r^e \tag{4.10}$$

4.2.2 Group-Target Tracking Gates

The tracking gates of group-targets are used for guaranteeing the falling of equivalent measurement into tracking gates with certain probability, but not for the filtering of valid returns. Thus, we need to set group-target association gates to realize the filtering of valid measurements. Since random errors are included in radar measurements, we can only guarantee target members to fall inside association gates in the next cycle, considering the geometric area of the steady groups validated in this cycle together with random errors of the marginal measurements of steady groups. According to Eq. 3.27 in Chap. 3, we can directly obtain the spatial geometric area of steady groups at time $k - 1$:

$$V(k-1) = D_1(k-1) \times D_2(k-1) \times \cdots \times D_I(k-1) \tag{4.11}$$

In Eq. 4.11, I is the dimension of a steady group.

When not considering group-target combination and splitting, since group-targets have mean cinematical behavior, namely location-displacement features, to some extent, the moving state of equivalent measurement approximately represents the state of every measurement and the variance of every measurement is less than the variance of equivalent measurement. Thus, assuring the correct reception probability coefficient of equivalent measurement can also assure that of every measurement. Furthermore, assuring the correct reception probability of marginal measurement can also assure that of non-marginal measurement. The variance of equivalent measurement is definitely greater than the variance of single measurement, so that for convenience of practical implementation, we use the variance of equivalent measurement to substitute for the variance of single measurement. According to group-target association area and tracking gate, the threshold of group-target association gates is:

$$\gamma_{ai}(k) = \frac{1}{2}D_i(k-1) + \gamma_T(k) \qquad (4.12)$$

So the volume of group-target association area at time k is:

$$V(k) = [D_1(k-1) + 2K_G\sigma_o] \times [D_2(k-1) \\ + 2K_G\sigma_o] \times \cdots \times [D_I(k-1) + 2K_G\sigma_o] \qquad (4.13)$$

Likewise, we can keep the consistency between group-target association area and group scale. Noticeably, although we use the variance of equivalent measurement to substitute for the variance of single measurement for convenience of practical implementation, this does not mean that group-target geometric area and tracking gate add up to the association gate in the next cycle. It actually represents the sum of group-target geometric association area and the variance of the K_G-fold measurement for assuring the correct reception probability of marginal measurements. The only difference is in that the variance of single measurement is substituted by the variance of equivalent measurement.

However, the function of group-target association gates necessarily leads to the problem of group-target combination and splitting. Since the target distance principle is the basis of group-target splitting and combination, according to the target distance given in Eq. 3.25, the final expression of association gates is:

$$\gamma_{Ai}(k) = \frac{1}{2}D_i(k-1) + K_t(k) + \gamma_T(k) \qquad (4.14)$$

This is the equation for computing group-target association gates.

We need to elaborate on the problem of association threshold of identifiable single targets. The precondition for setting single-target tracking threshold is that there are only tracking gates instead of association gates. Therefore, we have two options.

One is that we can directly adopt single-target tracking gates. According to Eq. 4.1, the tracking threshold is:

$$\gamma_{1i} = K_G\sigma_{ri} \qquad (4.15)$$

This method has high tracking precision. But it needs to redo splitting detection and group initiation when several single targets satisfy the target distance principle and they are going to merge.

The other is to introduce the group-target association gate formation method to single-target tracking gate. Namely consider target distance besides normal tracking gates. Then, we get:

$$\gamma_{2i} = K_G \sigma_{ri} + K_t(k) \tag{4.16}$$

This method is very simple. When multiple single targets satisfy the target distance principle, the combination can directly carry out. Its demerit lies in false alarm increases and precision degrades due to the man-made increase of tracking gates. The final aim of group-target tracking is to realize single-target tracking as early as possible. Therefore, this book chooses Eq. 4.15 to be the single-target tracking threshold.

4.3 The Nearest-Neighboring and All-Neighboring Association Algorithm for Single-Group-Target

For single-target probability data association algorithm, the number of true measurement is at most 1 and the rest are false measurements. This enables the computation of posterior probability of each return within tracking gates. But for group-targets, we cannot know precisely the number of real measurements that are derived from targets, and thus, we also cannot know the number of false measurements. Meanwhile, due to the non-one-to-one correspondence between targets and measurements, though we know the number of targets, we cannot distinguish what measurements are real measurements derived from targets and what measurements are false measurements derived from noise waves (this is why we need restrain noises in splitting detection, which will be discussed in detail). However, the all-neighboring algorithm assumes generally that the probability density of real measurements obeys Gauss distribution (also the basis of second-optimal Bayesian estimation) and noise waves obey uniform distribution. The probability data association algorithm is used for tracking single-targets of multi-returns, in which the tracks belong to real targets and equivalent returns are used to update filter state. The group-targets data association algorithm uses equivalent measurement to update filter state, and the tracks belong to equivalent measurement (This is why we adopt double thresholds. Association gates are used for validating group-target members and obtaining equivalent measurement, while tracking gates are used for validating group-target equivalent measurement.). Therefore, under condition of indistinguishability of multi-targets real measurements from false measurements of noise waves, we can only treat equally all the measurements within group-target association gates. So we need to find an expression to reflect the two different distributions and describe the contribution of both real measurements and false measurements on equivalent measurement.

The idea of the nearest-neighboring and all-neighboring data association algorithm is as follows. We combine NNDA (for its simplicity) and PDA (for its comprehensive consideration of each measurement, namely spatial accumulation) association algorithms to find out the essential connection of real measurements, false measurements, and association gates to the standard variance of filtering residual. And unite all these to compute the weights of each measurement within association gates. We get equivalent measurement according to the all-neighboring idea of PDA algorithm and realize single-group-target data association and track maintenance with equivalent measurement.

4.3.1   Computation of Group-Target Measurement Weights

As discussed before, the nearest-neighboring principle tells that the measurement nearest to the predicted center is regarded as real measurements derived from targets. Probability data association principle tells that every measurement within tracking gates is possibly derived from targets, but the probability density of each return is different [1, 16]. According to Eqs. 2.69 and 2.71, association probability $\beta_i(k)$ is directly proportional to $\exp\left[-\frac{1}{2}v_i^T S^{-1}(k) v_i(k)\right]$, which means the probability value $v_i(k)$ that the ith measurement is deviated from tracks. Parameter b depicts the probability with no correct measurements. According to Eqs. 2.70 and 2.72, $\beta_0(k)$ is also directly proportional to parameter b, and the denominator depicts the probability approximation of all possible events. Thus, combination coefficient $\beta_i(k)$ is a decreasing function of $v_i(k)$. In other words, the bigger the deviation of some measurement is, the smaller its probability is, and the smaller it makes contribution on equivalent measurement. This physical meaning is consistent with the nearest-neighboring association algorithm. For false measurements obeying uniform distribution, the probability that they fall inside association gates is generally a constant, which does not mean combination coefficient $\beta_0(k)$ is a constant. We find an interesting phenomenon according to Eqs. 2.71 and 2.72 that combination coefficients $\beta_i(k)$ and $\beta_0(k)$ are increasing functions of residual covariance $S(k)$. With a certain deviation between some observation data and tracks, the bigger $S(k)$ is, the bigger $\exp\left[-\frac{1}{2}v_i^T S^{-1}(k) v_i(k)\right]$ and b are. Directly speaking, $S(k)$ becomes bigger as the statistical deviation of this observation data from tracks become smaller. According to Eq. 4.1, for assuring correct reception probability, association threshold is a multiple of residual covariance. Namely the bigger the residual covariance is, the bigger the association threshold is. That is to say, residual covariance has consistent influence on real measurements and false measurements. The bigger the residual covariance is, the bigger $\exp\left[-\frac{1}{2}v_i^T S^{-1}(k) v_i(k)\right]$ and b of real and false measurements are. The weight of each measurement is directly relevant to association threshold. In short, the deviation of observation data from tracks and the association gate dimension based on standard variance of residual have uniform contribution on equivalent measurement as each measurement within association gates. So once we find an expression which can depict the consistent contribution made by the deviation of PDA algorithm and the standard variance of residual on equivalent measurement and reflect this kind of contribution at the same time, we can get the weight of each measurements within association gates by dividing the residual norm of each measurement by association gate threshold, and perfectly unify the nearest-neighboring algorithm and probability data association algorithm.

The association threshold is $\gamma_{Ai}(k)$ according to Eq. 4.14. The nearest-neighboring algorithm originally selects radar returns by checking whether the residual vector norm of measurements within association gates is less than $\gamma_{Ai}(k)$. If $g_i(k) \leq \gamma_{Ai}(k)$, the measurement is a candidate return [2, 68]. The residual is $v(k) = Y(k) - H(k)\widehat{X}(k/k-1)$, we use its covariance to perform normalization and construct the norm as:

$$
\begin{aligned}
g_i(k) &= \left\| Y_i(k) - H(k)\widehat{X}(k/k-1) \right\|^2 \\
&= \left[Y_i(k) - H(k)\widehat{X}(k/k-1) \right]^T S^{-1}(k) \cdot \left[Y_i(k) - H(k)\widehat{X}(k/k-1) \right] \\
&= v_i^T(k) S^{-1}(k) v_i(k)
\end{aligned}
$$

$$(4.17)$$

Then, we get:

$$\eta_i(k) = g_i(k)/\gamma_{Ai}(k) \tag{4.18}$$

η_i increases with $g_i(k)$, which means the longer the distance between returns and predicted center is, the smaller the posterior probability is. So the weight of the ith return is defined as:

$$\rho_i(k) = 1 - \eta_i(k) = 1 - \eta_i = 1 - g_i(k)/\gamma_{Ai}(k) \tag{4.19}$$

This is the equation for computing the weights of measurements based on residual norm.

4.3.2 Single-Group-Target Data Association Algorithm

As we know, traditional multi-target tracking gates, as a subspace in surveillance space, have two basic functions. First, restrict the number of target returns upon which we cannot make decision on. Second, assure track updating, tracking maintenance. The dimension of traditional multi-target tracking gates is decided by the probability of receiving correct returns. The concentric double-threshold association algorithm for group-targets also needs to take into account the probability of receiving correct returns, but here correct returns mean equivalent measurements instead of some concrete measurement. Association gates is far bigger than tracking gates, namely they are far bigger than the discrete space of equivalent measurement. If we only adopt this threshold, the probability of false-alarm occurrence is affirmatively high. Thus, in order to assure the probability of correct reception of equivalent measurement and the validation of group-target members, we adopt a concentric double-threshold association algorithm in group-target tracking. Its essence is to implement the two functions of multi-target tracking gate in two phases. The group-target association gate is used to determine the number of possible target returns, while the tracking gate is used to assure the correct reception of equivalent measurement. The two gates are concentric but with different sizes. The size of association gate is decided by group scale and changes with radar viewing angles no matter group-target have maneuver or not. The size of group-target tracking gate is decided by the probability of correctly receiving equivalent measurement, and its change in size is decided by whether group-target has maneuver and the maneuver model, etc. Association gates are regarded as constants when group scale does not change. The two concentric gates are not isolated. Their connecting tie lies in firstly the common center, secondly the product of the tracking gate constant for assuring correct reception probability in this cycle and the variance of marginal measurements, and thirdly the size of association gates in the next cycle which is decided by the geometric dimensions of group scales in this cycle. In short, group-target association gates are used to validate group-target members and form equivalent measurement, while tracking gates are used to validate equivalent measurement. Equivalent measurement is a false measurement based on association gate principles (a set with two or more targets can be called group-target) and will be processed only after all returns in this cycle have been processed. So tracking gate is a neat space with no noise wave and considers only the problem whether equivalent measurement falls inside. One of the important hypotheses of the probability data association algorithm is that there is at most a real measurement within tracking gates and targets and measurements are one-to-one corresponding, so we must consider the situation when all returns within tracking gates are all false measurements. The hypothesis of group-target association algorithm is that there are at least two targets within association gates, and targets and returns are non-one-to-one corresponding, so that the situation will not happen when all returns within tracking gates are all false measurements. Thus, we revise Eq. 4.19 and consider the completeness to get the group-target association probability:

$$\beta_i(k) = \frac{\rho_i(k)}{\sum_{i=1}^{m_k} \rho_i(k)}, \quad i = 1, 2, \ldots, m_k \tag{4.20}$$

According to Eqs. 4.19 and 4.20, with a certain measurement residual, the covariance becomes bigger means that $\gamma_{Ai}(k)$ becomes bigger, then η_i becomes smaller, and consequently $\rho_i(k)$ becomes bigger, vice versa. Similarly, with a certain residual covariance $\gamma_{Ai}(k)$, the residual becomes smaller, then η_i becomes smaller, and consequently $\rho_i(k)$ becomes bigger. We validate that this algorithm and PDA algorithm have consistency in computing the weights of each measurement, which means that this expression based on nearest neighboring and all-neighboring, as same as PAD algorithm can reflect the contribution of measurement deviation and the standard variance of residual on equivalent measurement. We all know that every algorithm has its cost, and the key is whether the cost is bearable or allowable. The cost of the measurement-weight computation method based on residual norm lies in that the weights have errors compared to the weights by PDA algorithm. The consideration of the errors is meaningful when applying the method to compute the equivalent measurement of multi-returns within single-target association gates. However, the errors are allowable or even need not to be considered in our case. Meanwhile, the weights assigned to each measurement are used only for computing equivalent measurement instead of constructing tracks of real measurements performed in PDA algorithm, the real and false measurements in this algorithm all have good consistency with those in PAD algorithm, and the measurements are normalized. Therefore, it is meaningless to consider the influence of these errors on equivalent measurement computation. PDA algorithm itself hopes to make equivalent measurement be more close to real measurement, while the measurement-weight computation method based on residual norm comprehensively takes into account the location information of each measurement within association gates and the size of association gate. Namely the group-target tracking algorithm itself hopes to obtain the real centroid of all measurements within association gates. In terms of the mathematical meaning of centroid, the measurement-weight computation method based on residual norm has more scientificalness rationality in computing equivalent measurement of group-targets.

In fact, $\rho_i(k)$ can be used as member weights directly under condition of stable flying state of group-targets and high signal-noise ratio. But the space partition based on weights of $\rho_i(k)$ has no completeness. Namely,

$$\sum_{i=1}^{m_k} \rho_i(k) \neq 1 \tag{4.21}$$

So the center of equivalent measurement has great fluctuations and group-targets have fake maneuver, when combination and splitting or sheltering of group-target members occur. Therefore, the method is not suitable for this situation.

According to the abovementioned equations, we easily get the state approximation of group-targets:

$$\widehat{X}(k/k) = \widehat{X}(k/k-1) + K_0^e(k)\left[\sum_{i=1}^{m_k} \beta_i(k)Y_i(k) - H(k)\widehat{X}(k/k-1)\right] \tag{4.22}$$

The equivalent measurement is:

$$Y_0^e(k) = \sum_{i=1}^{m_k} \beta_i(k) \, Y_i(k) \tag{4.23}$$

The equivalent gain is:

$$K_0^e(k) = P(k/k-1) H^T(k) \left[H(k) P(k/k-1) H^T(k) + R_0^e(k) \right]^{-1} \tag{4.24}$$

In Eq. 4.24:

$$R_0^e(k) = H(k) P(k/k-1) H^T(k) + \sum_{i=1}^{m_k} \beta_i^2(k) \, \sigma_u(k) \tag{4.25}$$

The one-step prediction equation is:

$$\widehat{X}(k+1/k) = \Phi(k+1/k)\widehat{X}(k/k-1) + K_0^e(k) \left[Y_0^e(k) - H(k)\widehat{X}(K/K-1) \right] \tag{4.26}$$

Equations 4.22–4.26 are the main equation set based on Kalman filtering for group-target tracking.

4.4   Summary

First, we define and design the group-target double gates of association gates and tracking gates. Based on the formation of association gates and tracking gates, the main difficulty for data association in single-group-target tracking is to get the weights of each measurement within association gates and to compute equivalent measurement. Although probability data association algorithm can compute the weight of each measurement within tracking gates and use the obtained equivalent measurement to update filters, it is designed for single-target tracking under condition of dense multi-returns, and thus cannot completely meet the requirement of group-target data association. Therefore, considering the feature of each measurement within tracking gates based on the simplicity of the nearest-neighboring association algorithm and the comprehensiveness of probability data association algorithm, we combine the two algorithms and propose **a single-group-target data association algorithm of nearest neighboring and all neighboring**. Taking the standard variance of filtering residual as a tie, this algorithm unifies the association probability of measurements with Gauss distribution and uniform distribution, and association gates. Under condition of multi–multi correspondence between targets and measurements, it equally deals with the measurements with different probability distributions within group-target association gates; constructs an uniform equation for computing weights of each measurement within association gates based on the nearest-neighboring idea; computes equivalent measurement based on the all-neighboring idea; and implements the data association and track maintenance of group-targets with equivalent measurement.

The simulation results prove the correctness and validity of the algorithm, as shown in Figs. 7.5, 7.6, 7.10, 7.11, and 7.16.

Chapter 5
Multi-Group-Target Data Association and Track Maintenance

5.1 Introduction

Multi-target tracking data association is a difficulty and also a key problem to be solved firstly in multi-target tracking. When tracking single target under situation of returns in density, the continuous occurrence of the measurement from near targets in the tracking gates causes continuous disturbance. The PDA algorithm, under condition of independent and consistent special distributions, treats all incorrect measurements as random disturbance models [19]. Therefore, when the existence of near targets causes incorrect disturbances to the models, not only the performance of target tracking will decrease significantly (continuous mistaken tracking occur), but also radar can easily lose targets. The joint probabilistic data association algorithm proposed by Bar-Shalom and his students [22, 28] sets the number of joint events as an exponential function of all candidate returns. Thus, the number of joint events, together with the mistaken association probability, increases exponentially with the density of returns, which causes overload to computers due to big computational request and degrade of the subsequent real-time processing capability. That is to say, although JPDA algorithm is applicable to multi-target tracking under condition of returns in high density, it is inapplicable to the tracking of multi-target in high density, which is already proved by the simulation result given in reference [1]. Meanwhile, many researchers have proposed improved algorithms of JPDA but also cannot solve satisfactorily the tracking problem of multi-target in high density.

The purpose of group-target tracking is to watch and track multi-targets in high density under condition of difficult identification or even unidentification of the high-density multi-targets. Noticeably, multi-targets with dense returns unnecessarily mean the high density of targets. In other words, multi-target tracking under condition of multi-returns in density and the tracking of high-density multi-target tracking are two different scenarios and conceptions [8, 10] with different focuses in implementation. JPDA algorithm aims to keep the correct tracking of each target

© National Defense Industry Press and Springer Science+Business Media Singapore 2017
W. Geng et al., *Group-target Tracking*, DOI 10.1007/978-981-10-1888-6_5

after track initiation. The group-target tracking algorithm aims to temporarily merge unidentifiable dense targets to group-targets instead of identifying them and turns to single-target tracking when identification becomes feasible with temporal and spatial changes. In fact, the group-target tracking algorithm adopts the tracking of group-target and single target during separate time periods and within separate sections, which equally transforms the problem of identification of dense targets to group tracking.

Adopting group-target tracking to realize dense multi-target tracking is also faced with the intersecting problem of association gates of multi-group-target. Based on the algorithm of single-group-target tracking association, regarding the multi–multi-correspondence between group-targets and measurements, we propose a multi-group-target data association algorithm imitating the idea of JPDA association algorithm. The idea of multi-group-target data association is as follows. (1) Detect all the group-targets with intersecting association gates. (2) Separate the common measurements within validly intersecting areas. (3) Compute the joint association probability of common measurements. (4) Normalize the weights of all the measurements within group-target association gates. (5) Synthesize them to get equivalent measurement.

This chapter proposes the idea of multi-group-target data association, based on JPDA association algorithm. We construct the splitting detection matrices of intersecting area detection within group-target association gates and common measurement within valid intersecting areas, define the joint events when group-target association gates intersection, and give the joint association probability algorithm and equivalent measurement algorithm under condition of multi-group-target association gates intersecting.

5.2   Idea of Double Multi–Multi-Correspondence of Multi-Group-Target Data Association

Multi-group-target tracking, as same as tradition multi-target tracking, is faced with association gate intersecting problem, as shown in Fig. 5.1. For research convenience, two hypotheses are given:

Hypothesis 1: Measurements within association gates are valid measurements, and targets and measurements have multi–multi-corresponding relationships.
Hypothesis 2: The all group-targets with valid association gates intersecting and the common measurement within intersecting areas have multi–multi-corresponding relationships.

According to JPDA association algorithm, the tracking gate is set to be consistent with the whole surveillance area in order to make the probability density function of each false measurement within the whole surveillance area be identical.

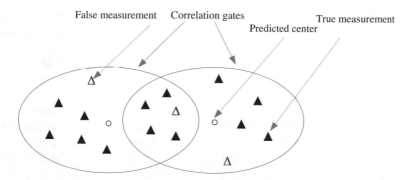

Fig. 5.1 Demonstration of association gate intersecting of multi-group-targets

Namely, JPDA algorithm deals with the probability density functions of true measurements and false measurements differently. However, as mentioned in Sect. 4.3 in Chap. 4, the group-target data association algorithm cannot distinguish between true measurements and false measurements within association gates; it can only regard all the measurement as valid measurements; namely, it treats true measurements and false measurements equally and the weights computation equation was given. In short, JPDA algorithm treats all tracking gates (same as the whole surveillance area) equally but treats true measurements and false measurements unequally; the group-target data association algorithm treats true measurements and false measurements equally (with same weights) but treats the association gates of each group-targets unequally. The aims of the two algorithms are both to obtain the weights of all measurements including false measurements, so the two have different methods but same aims. Since the independent dealing of false measurements does not exist under condition of group-target conception, it is obviously meaningless to treat tracking gates and surveillance area equally by using JPDA algorithm. In other words, even if the one-to-one correspondence between targets and measurements is satisfied, the different means in which JPDA and group-target data association algorithms deal with false measurements decide that the two algorithms deal with tracking gates and association gates differently. Therefore, association gates are kept for each group-target in the group-target data association algorithm.

An important reason for keeping association gates for each group-target is because there is double multi–multi-correspondence in group-target data association, which is new challenge ever. JPDA and data association algorithms [125, 126] are conditioned on the one-to-one and multi–multi-correspondence of targets and measurements, respectively. They both insist on the consistence between the tracking gate of each target and the surveillance area, to get the joint association probability. According to the above-mentioned two hypotheses, **the double multi–multi-correspondence of group-target data association means the multi–multi-correspondance between targets and measurements, and equivalent**

measurement of group-targets and common measurement within valid intersecting areas. Namely, equivalent measurement of group-targets and measurement have double multi–multi-corresponding relationships. The method to make consistency between the tracking gate and surveillance area and/or association gates and surveillance area of group-targets is difficult or even intractable. To get the joint association probability of under condition of double multi–multi-correspondence, we must distinguish the correspondence between targets and measurement from that between group-targets and common measurements. That is to say, all measurements belonging to each group-target need to be separated from surveillance area, and then, independent measurements inside association gates need to be separated from the common measurements inside intersecting areas.

By keeping association gates of each group-target, we can separate all measurements that belong to this group-target from surveillance area. Then, we can separate the independent measurements inside association gates from the common measurements. According to the equal processing of true measurements and false measurements in single-group-target data association algorithm, we process the independent measurements inside group-target association gates by using the method for multi–multi-correspondence of targets and measurements, which is consistent with the data association algorithm of single-group-target. Since group-target tracking gives only synthetic track instead of tracks of each target, whether it is multi–multi-correspondence or one-to-one correspondence between targets and measurements needs not to be considered. We only need to consider the multi–multi-correspondence problem between group-targets and common measurements, in order to multi-group-target data association under condition of double multi–multi-correspondence.

Based on the double multi–multi-correspondence between targets and measurements together with group-targets and common measurements, the idea of multi-group-target data association is discussed as follows.

We draw conclusions from splitting group detection and group initiation as follows. (1) The numbers of group-targets and/or single target are known. (2) After track initiation, the dimension of association gates of every track at time $k-1$ is also known. (3) The predicted location is known at time k; namely, the center of association gates is known. Based on the above-mentioned conclusions, we elaborate on the idea of multi-group-target data association according to the data association of JPDA algorithm. (1) Screen out all measurement inside group-target association gates according to association gate principles and take all returns inside association gates as valid measurements. (2) Taking the spatially statistical distances between the predicted centers of every pair of multi-group-targets as elements, construct the detection matrix of association gate intersecting and judge the intersecting of association gates based on the threshold (the sum of association gates of two group-targets). If the distance is less than the sum of association gates, intersection of association gates occurs. (3) Select the valid measurement of a group-target (with the least number of targets to decrease computation cost) when

association gates intersect and take the spatial distance between the valid measurement and the predicted center of the rest intersecting group-target as elements to
construct splitting detection matrix of common measurements. Then, we can inspect
the valid intersecting situation of association gates between group-targets and
determine which measurements are inside the intersecting areas (to compute the
number of common measurements). (4) Process the splitting detection matrix of
common measurements, remove the rows and columns with zero 1-norm to get
group-targets possessing valid intersecting, and separate all the measurement inside
association gates into two groups of common measurements and independent
measurements. The aim is to eliminate the number of false-intersecting groups
when no measurements fall into intersecting areas though association gates intersect, decrease unnecessary computation, and realize the separation of common
measurements and independent measurements. (5) Based on the intersecting
detection matrix of association gates and the splitting detection matrix of common
measurements, categorize multi-group-targets and all measurements into two sets:
the group-targets with valid intersecting of association gates and the common
measurements. Then, define association events and joint association events and
compute the joint association probability of common measurements. (6) Under
condition of multi–multi-correspondence between targets and measurements, process independent measurements with diverse probability distribution inside association gates equally to compute the association probability of independent
measurements. (7) Normalize the joint association probability of common measurements and the association probability of independent measurements to get the
weight of each measurement and compute equivalent measurement and finally
implement the correct association and track maintenance of multi-group-targets.

Because the multi-group-target data association algorithm does not take the
whole surveillance area as the association area and it keeps the association gate of
each group-target, we do not need to perform association judgment on all returns
and all group-targets within the whole surveillance area. The valid intersecting
judgment based on the splitting detection matrix of common measurements eliminates invalid intersecting group-targets, avoids unnecessary dealing with all the
group-targets with intersecting association gates, eliminates the influence of
non-common measurements (with zero 1-norm), and eventually decreases the
number of joint events. Thus, computation cost is decreased significantly.

5.3   Construction of Intersecting Detection Matrix of Multi-Group-Target Association Gates

According to the analysis in the last section, JPDA algorithm is based on two
hypotheses. First, one measurement can have one source, namely that a measurement is derived from a target, a noise wave, or a false alarm. Second, for a given

target, at most one measurement is derived from it. Namely, if a target generates multiple measurements, only one of them is regarded as true and the rest are regarded as false. In fact, due to the close distance between members of group-targets and the different angle of view of radar, there exists the situation accidentally when one target is corresponding to multiple measurements and/or multi-targets are corresponding to one or several measurements. For example, the number of returns from a group-target flying in formation is relevant to the elements of angle, direction, and polarization of the formation, so we can hardly make the number of returns correspond with the number of targets one by one. Due to the factors of sheltering between members in this formation and the subsequent indistinguishability, there exists the situation when one measurement is corresponding to multiple targets. JPDA algorithm needs to assume that the number of targets is known [1, 22]. Although the number of targets inside each group-target is not always known under the condition of group-target tracking, the number of group-targets inside surveillance space is known. Therefore, from the perspective of tracking group-targets as independent targets, JPDA algorithm is consistent with group-target tracking algorithm.

After going through splitting group detection and group initiation, the total of the measurement of the kth scan within surveillance areas is m_k, under the condition that tracks of multi-group-target have been initiated. Splitting group detection categorizes it into G subgroups, and the valid measurement of each subgroup is $m_1, m_2, \ldots, m_j, \ldots, m_G$. They satisfy:

$$m_k = m_1 + m_2 + \cdots + m_j + \cdots + m_G \tag{5.1}$$

The valid measurement of the jth subgroup at time k is $Y_{j_i}(k)(i = 1, 2, \ldots, m_j; j = 1, 2, \ldots, G)$. The equivalent measurement got from any subgroup is $Z_j(k)$, and the predicted center is $Z_j(k/k - 1)$.

Definition 1 The spatial distance between the jth subgroup and the predicted center of the equivalent measurements of any other subgroup l is defined as:

$$d_{jl}(k) = d_{lj}(k) = |Z_j(k/k - 1) - Z_l(k/k - 1)| \tag{5.2}$$

When $d_{jl}(k) = d_{lj}(k) < \gamma_j + \gamma_l$, subgroup j and subgroup l intersect, $\gamma_j \leftrightarrows \gamma_l$ are the association thresholds of the corresponding subgroups, and $j \neq l$. Let

$$\Delta_{jl}(k) = \begin{cases} 1, & d_{jl}(k) \leq \gamma_j + \gamma_l \\ 0, & d_{jl}(k) > \gamma_j + \gamma_l \end{cases} \tag{5.3}$$

Since $\Delta_{jl} = \Delta_{lj}$, the combination number of subgroups for pair-intersecting judgments is C_G^2. The intersecting detection matrix of association gates is as follows:

$$
\begin{bmatrix}
 & Z_1(k) & Z_2(k) & \cdots & Z_{(G-1)}(k) & Z_G(k) \\
Z_1(k) & 1 & \Delta_{12}(k) & \cdots & \Delta_{1(G-1)}(k) & \Delta_{1G}(k) \\
Z_2(k) & 0 & 1 & \cdots & \Delta_{2(G-1)}(k) & \Delta_{2G}(k) \\
\vdots & \vdots & \vdots & \cdots & \vdots & \vdots \\
Z_{(G-1)}(k) & 0 & 0 & \cdots & 1 & \Delta_{(G-1)G}(k) \\
Z_G(k) & 0 & 0 & \cdots & 0 & 1
\end{bmatrix}
\tag{5.4}
$$

When $\Delta_{jl}(k) = 1$, group-targets j and l intersect. When the judgment matrix is a unit matrix, there is no intersecting between every two group-targets, so joint association computation is unnecessary. After obtaining the intersecting situation of group-targets, we will next study splitting the algorithm of common measurement.

5.4 Construction of Splitting Detection Matrix of Multi-Group-Target Common Measurements

Definition 2 d_{jil} is the distance between the valid measurement j_i of the jth subgroup and the predicted center of any subgroup l. When $d_{jil}(k) \le \gamma_l$, a valid measurement j_i occurs within the intersecting area of subgroups j and l, namely the common measurement. γ_l is the association threshold of subgroup l. Let

$$
\varepsilon_{jil}(k) = \begin{cases} 1, & d_{jil}(k) \le \gamma_l \\ 0, & d_{jil}(k) > \gamma_l \end{cases}
\tag{5.5}
$$

According to Eq. 5.34, the group-targets with intersections are already obtained; next, the splitting detection matrix will be constructed when the valid measurement j_i of the jth group-target falls inside the association gates of other group-targets.

Assume that the worst association gates intersecting happens, namely association gates of all group-targets intersect. We remove the jth group-target to eliminate the redundant computation for the same group-target and get the splitting matrix of common measurements when the measurement of group-target j falls inside the intersecting areas of other groups:

$$
\begin{bmatrix}
 & Z_1(k) & Z_2(k) & \cdots & Z_l(k) & \cdots & Z_{G-1}(k) \\
Y_{j_1}(k) & \varepsilon_{j_1 1}(k) & \varepsilon_{j_1 2}(k) & \cdots & \varepsilon_{j_1 l}(k) & \cdots & \varepsilon_{j_1 G-1}(k) \\
Y_{j_2}(k) & \varepsilon_{j_2 1}(k) & \varepsilon_{j_2 2}(k) & \cdots & \varepsilon_{j_2 l}(k) & \cdots & \varepsilon_{j_2 G-1}(k) \\
\vdots & \vdots & \vdots & \cdots & \vdots & \cdots & \vdots \\
Y_{j_{m_j}}(k) & \varepsilon_{j_{m_j} 1}(k) & \varepsilon_{j_{m_j} 2}(k) & \cdots & \varepsilon_{j_{m_j} l}(k) & \cdots & \varepsilon_{j_{m_j} G-1}(k)
\end{bmatrix}
\tag{5.6}
$$

According to Eq. 5.6, when the 1-norm of columns is zero, there is no common measurement between group-targets l and j. Namely, the association gates of group-targets l and j do not intersect, or there is no valid measurement falling inside the intersecting areas when intersection happens. In this case, the group-targets l and j are independent from each other, and they are called invalid intersecting group-target. If the 1-norm of a column is not zero, the value of its 1-norm indicates the number of group-targets within which a given measurement of group-target l is located; namely, the number of the rest group-targets with which group-target l intersects. The 1-norm expression of the l th column is as follows:

$$n_{q_l} = \sum_{j=1}^{m_j} \left| \varepsilon_{j_i l} \right| \tag{5.7}$$

Denote the columns with zero 1-norm by n_{q_l0}. When group-target l has no valid intersecting with any group-targets, the total of the (independent) measurements is as follows:

$$n_{q0} = \sum_{\substack{l=1 \\ d_{j_i l} > \gamma_l}}^{G-1} n_{q_{j_i}0} \tag{5.8}$$

Then, we get the number of common measurements, $m_j - n_{q0}$.

When the 1-norm of rows is zero, there is no intersecting between group-target l and any group-targets. Thus, the valid measurement only belongs to group-target l, and the group-target has no intersecting or has only invalid intersecting. The value of 1-norm indicates the number of valid measurement of group-target j that falls within group-target l. The 1-norm of the j_i row is as follows:

$$n_{p_{j_i}} = \sum_{l=1}^{G-1} \left| \varepsilon_{j_i l} \right| \tag{5.9}$$

Denote the row with zero 1-norm by n_{q_l0}. The number of group-targets inside which none of independent measurement falls is as follows:

$$n_{p0} = \sum_{\substack{j=1 \\ d_{j_i l} > \gamma_l}}^{m_j} n_{p_{j_i}0} \tag{5.10}$$

n_{p0} is the number of group-targets which have no intersecting or have only invalid intersecting. Thus, the number of group-targets which got at least one measurement inside is $m_p = (G-1) - n_{p0}$.

As for the jth group-target, the number of relevant valid measurements is m_j, the number of common measurements is $m_p = m_j - n_{q0}$, and the number of independent measurements which belong to itself is n_{q0}. Likewise, if there are G group-targets within surveillance area, the maximum number of group-targets which share common measurement is $(G - 1) - n_{q0}$. The contribution of n_{q0} independent measurements on equivalent measurement of group-targets can be calculated by directly using the single-target association probability algorithm, but the common measurements within validly intersecting areas need to be calculated by joint association algorithm. Finally, we can process the joint association between group-target l and other group-targets in parallel, to finish all the computation of joint association probability of common measurements within all valid intersecting group-targets. Similarly, the number of all validly intersecting group-targets can be obtained by processing the association between the measurement of group-target l and other group-targets. Next, we will construct the equation to compute the common-measurement association probability within validly intersecting areas.

5.5 Computation of Multi-Group-Target Combination Association Probability and Equivalent Measurement

According to the hypotheses in Sect. 5.2, group-target joint association events incorporate two sets. The first is the group-target set based on validly intersecting group-targets, and the second is the measurement set based on common measurements. That is to say, group-targets have common measurements when they have valid intersecting association gates, and each common measurement is definitely derived from group-targets with valid intersection. After finishing the partition of this kind of association events, we adopt Bayesian principles to get their association probability, respectively. Finally, we add up the association probabilities of the two sets to get the joint association probability of common measurements.

According to references [68-69], under condition of multi–multi-correspondence between group-targets and common measurements, we process group-targets on a basis of group-targets through common-measurement splitting detection matrix, based on the hypothesis that "group-targets all have common measurements when association gates have valid intersecting." This processing does not consider the condition that "every common measurement is definitely derived from group-targets." Likewise, we process common measurements on a basis of common measurement through common-measurement splitting detection matrix, based on the hypothesis that "every common measurement is definitely derived from group-targets." And this processing does not consider the condition that "group-targets all have common measurements when association gates have valid intersections." This processing method is preconditioned on the repeated usage of group-targets with common measurements and validly intersecting association gates. The difference between the method and JPDA algorithm lies in the following: (1) Measurement in JPDA algorithm is corresponding to targets one by one; namely, measurement and targets cannot be repeatedly used; (2) common measurements can be repeatedly used on a basis of group-target; namely, under the restriction of common-measurement splitting matrix, group-targets all have common measurements when association gates have valid intersections, which solve the problem of "one common measurement is associated to multi-group-target." In contrary, group-targets can be repeatedly used on a basis of common measurement, which solves the problem of "one group-target with valid intersection is associated to multiple common measurements." This way, we unite the problems of "one common measurement is associated to multi-group-target with valid intersections" and "one group-target is associated to multiple common measurements" and realize the multi–multi-correspondence between common measurements and group-targets with valid intersection of association gates.

As for the processing of false measurements, JPDA algorithm adopts the strategy of one-to-one correspondence between measurements and targets by using measurement indicator and target detection indicator. However, it cannot realize the one-to-one correspondence between measurements and equivalent measurement of group-targets in group-target data association, because the number of false measurements is hard to be known. Since the equivalent measurement is corresponding to the real target in JPDA algorithm and every equivalent measurement has multiple measurements in a real sense, we treat equally both real measurements and false measurement and adopt the noise-restraining algorithm during the course of combination and splitting detection to solve the complicated processing of false measurement in JPDA algorithm.

JPDA and GPDA (general probability data association), as multi-target tracking algorithms, are discussed in references [1, 68–69]. They do not set independent association gates for each target, but only regard the association gate of each target as being consistent with the whole surveillance area. The multi-group-target association algorithm proposed in this book is based on the fact that each group-target has its own association gate. Thus, we give the following definitions.

Definition 3 $\Lambda_j(k) = [\delta_{pq}(k)], p = 0, 1, \ldots, m_p; q = 0, 1, \ldots, G_q$ is the association probability matrix in a sense of nearest neighboring, where Λ_j is the association probability matrix between the valid measurements within the validly intersecting area of the jth group-target and the other group-targets, and δ_{pq} is the weight between valid measurement p and group-target q. According to Eq. 4.17 in Chap. 4, we directly give:

$$\delta_{pq}(k) = 1 - \frac{d_{pq}(k)}{\gamma_q} \tag{5.11}$$

where γ_q is the association threshold of group-target q. The association probability matrix is as follows:

$$\begin{bmatrix} & Z_j(k) & Z_1(k) & Z_2(k) & \cdots & Z_{G_q-1}(k) \\ Y_1(k) & 1 & \delta_{11}(k) & \delta_{12}(k) & \cdots & \delta_{1G_q-1}(k) \\ Y_2(k) & 1 & \delta_{21}(k) & \delta_{22}(k) & \cdots & \delta_{2G_q-1}(k) \\ \vdots & \vdots & \vdots & \vdots & \cdots & \vdots \\ Y_{m_p}(k) & 1 & \delta_{m_p1}(k) & \delta_{m_p2}(k) & \cdots & \delta_{m_pG_q-1}(k) \end{bmatrix} \qquad (5.12)$$

Definition 4 Γ_j is the joint association event that satisfies the multi–multi-correspondence between measurements and targets. Γ_{m_p} is the joint association event that the measurements within validly intersecting areas are likely derived from equivalent measurement of multi-group-targets. Γ_{G_q} is the joint association event that the equivalent measurement of group-targets is likely derived from valid measurements within multiple validly intersecting areas. Then, we get:

$$\Gamma_j = \Gamma_{m_p} \bigcup \Gamma_{G_q} \qquad (5.13)$$

where $\Gamma_j = \{\theta_j(k)\}$.

Assume that $\Gamma_{m_p} = \{\theta_{p_i}(k)\}_{p_i=1}^{n_{p_k}}$ denotes the set of all possible association events on a basis of measurement at time k, and $n_{p_k} = m_p^2 \cdot (G_q+1)^2$ denotes the number of joint association events in Γ_{m_p}.

$$\theta_{p_i}(k) = \bigcap_{p=1}^{m_p} \theta_{pq}^{p_i}(k) \qquad (5.14)$$

$\theta_{p_i}(k)$ denotes the p_ith joint association event. $\theta_{pq}^{p_i}(k)$ denotes the association event that measurement p is derived from group-target q in the p_ith joint association event.

Assume that $\theta_{yj}(k)$ denotes the event that the yth measurement and group-target j associate.

$$\theta_{yi}(k) = \bigcup_{p_i}^{n_{pk}} \theta_{pq}^{p_i}(k) \qquad (5.15)$$

Likewise, we can get the set of all possible association events on a basis of group-target, $\Gamma_{G_q} = \{\theta_{q_j}(k)\}_{q_j=1}^{n_{q_k}}$. $n_{q_k} = G_q^2 \cdot (m_p+1)^2$ denotes the number of joint association events in Γ_{G_q} where

$$\theta_{q_j}(k) = \bigcap_{q=1}^{G_q} \theta_{pq}^{q_j}(k) \qquad (5.16)$$

$\theta_{q_j}(k)$ denotes the q_jth joint association event. $\theta_{pq}^{p_i}(k)$ denotes the association event that group-target q is derived from measurement p in the q_jth joint association event.

$$\theta_{jy}(k) = \bigcup_{q_i}^{n_{qk}} \theta_{pq}^{q_j}(k) \tag{5.17}$$

$\theta_{jy}(k)$ denotes the event that group-target j and the yth measurement associate.

In order to guarantee completeness, we perform normalization to the normalized probability matrices on a basis of measurements and group-targets, respectively. Normalizing the association probability matrix on a basis of common measurement, we get the normalized matrix $E_Y = [\xi_{pq}]$, where

$$\xi_{pq} = \frac{\delta_{pq}}{c_p} \tag{5.18}$$

In Eq. 5.18, c_p is the normalization coefficient on a basis of group-target, where

$$c_p = \sum_{p=1}^{m_p} \xi_{pq} \tag{5.19}$$

Similarly, normalizing the association probability matrix on a basis of group-target, we get the normalized matrix $E_Z = [\zeta_{pq}]$, where

$$\zeta_{pq} = \frac{\delta_{pq}}{c_q} \tag{5.20}$$

In Eq. 5.20, c_q is the normalization coefficient on a basis of group-target q, where

$$c_q = \sum_{q=1}^{m_G} \zeta_{pq} \tag{5.21}$$

According to Definition 4, the measurement association probability within intersecting areas is as follows:

$$\sum_{\Gamma_j} P[\theta_j(k)/Y^k] = \sum_{\Gamma_{mp}} P[\theta_{p_i}(k)/Y^k] \bigcup \sum_{\Gamma_{Gq}} P[\theta_{q_j}(k)/Z^k] \tag{5.22}$$

Based on Bayesian equation, the conditional probability of association events based on all measurements is as follows:

$$P[\theta_{p_i}(k)/Y^k] = P[\theta_{p_i}(k)/Y(k), Y^{k-1}] = P[\theta_{p_i}(k)/Y^{k-1}] \cdot P[Y(k)/\theta_{p_i}(k), Y^{k-1}]$$

(5.23)

where $P[\theta_{p_i}(k)/Y^{k-1}] = P[\theta_{p_i}(k)]$ is the prior probability. Compared with the concerned jth group-target, the yth measurement falls inside the association gate of group-target j, so its prior probability is as follows:

$$P[\theta_{p_i}(k)/Y^{k-1}] = P[\theta_{yj}] = \xi_{pj}$$

(5.24)

Since $P[Y_y(k)/\theta_{p_i}, Y^{k-1}] = P[Y_y(k)/\theta_{pq}^{p_i}, Y^{k-1}] = \xi_{pq}$, we get:

$$P[Y(k)/\theta_{pq}^{p_i}, Y^{k-1}] = \prod_{\substack{p=1 \\ q \neq j}}^{m_p} P[Y_y(k)/\theta_{pq}^{p_i}, Y^{k-1}] = \prod_{p=1}^{m_p} \xi_{pq}$$

(5.25)

Substituting Eqs. 5.24 and 5.25 into Eq. 5.23, we get:

$$P[\theta_{p_i}(k)/Y^k] = \xi_{yj} \prod_{\substack{p=1, q=1 \\ q \neq j, p \neq y}}^{m_p, G_q} \xi_{pq}$$

(5.26)

According to Eqs. 5.15 and 5.26, we get:

$$
\begin{aligned}
P[\theta_{yj}(k)/Y^k] &= P\left[\bigcup_{p_i=1}^{n_{pk}} \theta_{pq}^{p_i}(k)/Y^k\right] \\
&= \sum_{p_i=1}^{n_{pk}} P[\theta_{p_i}(k)/Y^k] = \sum_{p_i=1}^{n_{pk}} \xi_{yj} \prod_{\substack{p=1, q=1 \\ p \neq y, q \neq j}}^{m_p, G_q} \xi_{pq} = \xi_{yj} \prod_{\substack{p=1 \\ p \neq y}}^{m_p} \sum_{\substack{q=1 \\ q \neq j}}^{G_q} \xi_{pq}
\end{aligned}
$$

(5.27)

Similar to the above-mentioned process on a basis of measurements, we substitute the set of measurements Y^k with the set of group-targets Z^k and directly give the joint association probability on a basis of all group-targets:

$$
\begin{aligned}
P[\theta_{jy}(k)/Z^k] &= P\left[\bigcup_{q_i=1}^{n_{qk}} \theta_{pq}^{p_i}(k)/Z^k\right] \\
&= \sum_{q_i=1}^{n_{qk}} P[\theta_{q_i}(k)/Z^k] = \sum_{q_i=1}^{n_{qk}} \xi_{yj} \prod_{\substack{p=1, q=1 \\ p \neq y, q \neq j}}^{m_p, G_q} \zeta_{pq} = \zeta_{yj} \prod_{\substack{p=1 \\ p \neq y}}^{m_p} \sum_{\substack{q=1 \\ q \neq j}}^{G_q} \zeta_{pq}
\end{aligned}
$$

(5.28)

Substituting Eqs. 5.27 and 5.28 into Eq. 5.22, we get:

$$\sum_{\Gamma_j} P[\theta_j(k)/Y^k] = \xi_{pj} \prod_{p=1}^{m_p} \sum_{\substack{q=1 \\ q \neq j}}^{G_q} \xi_{pq} + \zeta_{pj} \prod_{\substack{q=1 \\ q \neq j}}^{G_q} \sum_{p=1}^{m_p} \zeta_{pq} \qquad (5.29)$$

In order to guarantee completeness, we normalize the group-target j and get the association probability of the measurement within its intersecting area:

$$\partial_{yj} = \frac{1}{c} \sum_{\Gamma_j} P[\theta_j(k)/Y^k] = \frac{1}{c} \left(\xi_{pj} \prod_{p=1}^{m_p} \sum_{\substack{q=1 \\ q \neq j}}^{G_q} \xi_{pq} + \zeta_{pj} \prod_{\substack{q=1 \\ q \neq j}}^{G_q} \sum_{p=1}^{m_p} \zeta_{pq} \right) \qquad (5.30)$$

In Eq. 5.30,

$$c = \sum_{p=1}^{m_p} \xi_{yj} \prod_{\substack{p=1 \\ p \neq y}}^{m_p} \sum_{\substack{q=1 \\ q \neq j}}^{G_q} \xi_{pq} + \sum_{q=1}^{G_q} \zeta_{yj} \prod_{\substack{q=1 \\ q \neq j}}^{G_q} \sum_{\substack{p=1 \\ p \neq y}}^{m_p} \zeta_{pq} \qquad (5.31)$$

c in Eq. 5.31 is the common normalization coefficient for group-targets with common measurements and non-common measurements.

According to Eqs. 5.24, 5.30, and 5.31, we normalize the all measurements within the association gate of group-target $k-1$, and the normalized probability of the measurement outside validly intersecting areas is as follows:

$$\beta_{ij} = \frac{\rho_i(k)}{\sum_{i=1}^{n_{p0}} \rho_i(k) + c}$$

$$= \frac{\rho_i(k)}{\sum_{i=1}^{n_{p0}} \rho_i(k) + \sum_{p=1}^{m_p} \xi_{yj} \prod_{\substack{p=1 \\ p \neq y}}^{m_p} \sum_{\substack{q=1 \\ q \neq j}}^{G_q} \xi_{pq} + \sum_{q=1}^{G_q} \zeta_{yj} \prod_{\substack{q=1 \\ q \neq j}}^{G_q} \sum_{\substack{p=1 \\ p \neq y}}^{m_p} \zeta_{pq}}$$

$$(5.32)$$

The normalized probability of the measurement inside validly intersecting areas is as follows:

$$\eta_{ij} = \frac{\sum_{\Gamma_j} P[\theta_j(k)/Y^k]}{\sum_{i=1}^{n_{p0}} \rho_i(k) + c}$$

$$= \frac{\xi_{pj} \prod_{p=1}^{m_p} \sum_{\substack{q=1 \\ q \neq j}}^{G_q} \xi_{pq} + \zeta_{pj} \prod_{\substack{q=1 \\ q \neq j}}^{G_q} \sum_{p=1}^{m_p} \zeta_{pq}}{\sum_{i=1}^{n_{p0}} \rho_i(k) + \sum_{p=1}^{m_p} \xi_{yj} \prod_{\substack{p=1 \\ p \neq y}}^{m_p} \sum_{\substack{q=1 \\ q \neq j}}^{G_q} \xi_{pq} + \sum_{q=1}^{G_q} \zeta_{yj} \prod_{\substack{q=1 \\ q \neq j}}^{G_q} \sum_{\substack{p=1 \\ p \neq y}}^{m_p} \zeta_{pq}}$$

$$(5.33)$$

Eventually, we get the equivalent measurement of group-target j:

$$Y_0^E(k) = \sum_{i=1}^{n_{p0}} \beta_{ij} Y_i(k) + \sum_{y=0}^{m_p} \eta_{ij} Y_y(k) \tag{5.34}$$

The Kalman filtering estimation equation, gain, and covariance computation for multi-group-targets are similar to those for single-group-targets. Here, we do not give unnecessary details again.

5.6   Summary

Borrowing the idea of JPDA algorithm for computing joint association probability, we break the restriction of the one-to-one correspondence between targets and measurements and propose a target association algorithm of multi-group-targets under the condition of double multi–multi-correspondence between targets and measurements, together with group-targets and common measurements. First, the algorithm separates members of group-targets from surveillance area according to association gates of group-targets. Second, taking the spatial distance between equivalent measurements of group-targets as the elements, construct the association gate intersection detection matrix and judge which group-targets will intersect according to the 2-norm between the elements. Third, construct a common-measurement splitting detection matrix and use the 1-norm of rows and columns to realize the common-measurement splitting detection on what group-targets validly intersect and how many measurements are within intersecting areas. Forth, categorize all group-targets and measurements into two sets: the group-targets with valid intersections and the common measurements and give the equation to compute the joint association probability of common measurements, taking multi–multi-correspondence of group-targets and common measurements as a marginal condition. Fifth, with the help of common-measurement detection matrix, categorize all measurement within association gates into independent measurements and common measurements and deal with the independent measurement with diverse probability distribution equally to compute the association probability of each independent measurement under condition of multi–multi-correspondence between targets and measurements. Finally, normalize the joint association probability of common measurements and the association probability of independent measurements outside intersecting areas to get the weight of each measurement of group-target, construct the equation for computing equivalent measurement based on the all-neighboring idea, and finally implement the data association and track maintenance of multi-group-targets.

The main differences between group-target data association algorithm and JPDA algorithm are as follows. (1) JPDA is an association algorithm under condition of one-to-one correspondence, while the multi-group-target algorithm is an association algorithm under condition of multi–multi-correspondence. (2) JPDA algorithm treats the whole surveillance area as association area, eliminates the tracking gate of single target, while the multi-group-target algorithm keeps the association gate of each group-target. (3) JPDA algorithm exerts itself to identify and correctly track every target, while the multi-group-target algorithm aims at the tracking of the whole targets in density and implements single-target tracking during separate time periods and within separate sections. (4) To some density extent, JPDA algorithm cannot distinguish, while the multi-group-target algorithm is not restricted by density. (5) JPDA treats all tracking gates (same as the whole surveillance area) equally but treats true measurements and false measurements unequally; meanwhile, the group-target data association algorithm treats all true measurements and false measurements equally (with same weights) but treats the association gates of each group-targets unequally. (6) JPDA algorithm is a multi-target data association algorithm under condition of one-to-one correspondence between targets and measurements, while the group-target data association algorithm is a data association algorithm under condition of double multi–multi-correspondence between targets and measurements, together with group-targets and common measurements. Therefore, the multi-group-target association algorithm is applicable more widely and has important value both in theory and in practice.

The simulation results prove the correctness and validity of the multi-group-target association algorithm, as shown in Figs. 7.5, 7.6, 7.12, 7.13–7.16.

Chapter 6
Detection of Group-Target Combination and Splitting and Situation Cognition

6.1 Introduction

Purely from the perspective of target tracking, we have completed the pretreatment of group-target measurement, group-target track initiation, single group-target data correlating and track maintenance, and multi-group-target data correlating and track maintenance. We simply need to conduct group-target track termination to finish all the necessary to group-target tracking. However, multi-target tracking computes the tracks of single targets, while group-target tracking computes the track of the whole group-target. Since considered as a whole, a group-target necessarily involves new member combination, existing member splitting, and the maneuver and changes of inner members. Furthermore, it may even involve the combination of another whole and the splitting of a local whole, which can give rise to the fluctuation of the group-target centroid. The fluctuation of the group-target centroid necessarily causes the maneuver-like phenomena of the group-target. So that group-target combination and splitting detection and subsequent group-target stable tracking need to be solved. Ostensibly group-target combination and splitting detection are to help solve group-target tracking, but the essential issue has changed from tracking problem at the data layer to the situation cognition problem at the cognition layer.

In fact, the group-target data correlating problem itself discussed in the previous chapters incorporates the group-target situation correlating problem. If we obtain the state of the inner members of a group-target (new member combination, exiting member departure, and inner member maneuver) as well as we obtain the whole characteristics of the group-target, we correspondingly are aware of the holistic state and spatial distribution of these multi-targets, together with the dynamic relationships between each pair of targets and the moving prediction of splitting targets. Obviously, the information from group-target tracking already incorporates the basic elements of threat estimation, situation association, and situation synthesizing.

© National Defense Industry Press and Springer Science+Business Media Singapore 2017
W. Geng et al., *Group-target Tracking*, DOI 10.1007/978-981-10-1888-6_6

The combination and splitting of a group-target depict not only the maneuver of the group-target but also the holistic situation. Therefore, the group-target combination and splitting detection belong to not only group-target maneuver detection but also situation cognition. This chapter describes the group-target maneuver first and then discusses the problems of group-target combination and splitting detection, and group termination.

6.2   Group-Target Maneuver Description

Actually for any tactical formation of planes, its formation flying is a process instead of a purpose. There is possibly combination of new members to the formation and member splitting when close to the attack objectives (of course false alarms and missing detections may occur) [7]. The combination and splitting of group members will cause changes in the group scale and the centroid, compared to group-target tracking [2]. This can also cause tracking error increases and even tracking failures similar to what happens in traditional target tracking. Thus, we define group-target maneuver as the group-target centroid fluctuation due to member combination and splitting. In other words, if acceleration changes characterize the maneuver of traditional target tracking, group scale changes, or centroid fluctuations characterize the maneuver of a group-target. The involving detection method is combination and splitting detection.

Till now, there are not many references studying the combination and splitting detection of group-targets. The commonly used detection is conducted according to location information. We further reveal that the group-target member splitting is related to velocity and direction besides location. Therefore, location, velocity, and direction are the three elements of the group-target member splitting detection. All the judgment principles of combination and splitting can be regarded as the logic combination of the three elements.

To help investigate the group-target combination and splitting problem, we define as follows.

Definition 1: The process of new measurements, which satisfy association gate principles, being combined to a given group-target or multi-group-targets, which satisfy group-target distance principles, being combined to a new group-target, is called group-target combination.

Definition 2: The process of group-target members departing from the previous group-target one by one or holistically and forming some group-targets and/or distinguishable single targets is called group-target splitting.

The basic idea of group-target combination and splitting is as follows.

As for the combination issue, first, consider the combination of new observation data with some group-targets. The observation data falling into the association gate are taken temporarily as the member of the group-target according to association gate principles, and then, a new group-target comes into being after a few cycles of validation. When the uncorrelated new observation data are confirmed to be target return instead of false alarms, new tracks will be initiated. Otherwise, the new data will be discarded. Namely, the combination of new observation data with some group-target has utilized the association combination principle of observation data and tracks. Second, consider the combination of group-targets with stable tracks and/or single-target group. Track–track association method is adopted to perform the combination determination and judge whether the combination is satisfied according to track direction and velocity information. In the meantime, track history is considered to avoid mistaken combination and subsequent batch mixture when tracks fork. This correspondingly introduces the distributed multi-sensor track–track combination algorithm to the group-target tracking combination.

As for the splitting issue, we propose an idea of presetting sub-wave gates within the association gates of group-targets to conduct double association. By judging whether the neighboring cycles-related measurements fall into the sub-wave gates, we determine whether group-target members split. And we process the weights of the measurements, which are inside the association gates and outside the sub-wave gates, by considering the directions and velocities. Finally, we implement splitting detection, taking into account comprehensively the location, velocities, and directions of the measurements inside the association gates.

The group-target combination and splitting detection, regarded as a key element in the group-target tracking, is essentially different from the splitting group detection and group initiation discussed in Chap. 3. They are two issues at different two stages of group-target tracking. The splitting group detection and group initiation is to combine the observation data sets, which satisfy the target distance principles, into transient groups and then into stable groups through the step-by-step validation of group initiation. In the meantime, it initiates the track of the group-target. The splitting group detection and group initiation lasts until all returns get stable tracks. Once the stable tracks of the group-targets begin, the next is to deal with the measurement pretreatment as group-target combination and splitting detection. The existing group-target has changed in scale, due to new member combination, old member splitting, and some group-targets' combination and splitting. The splitting group detection and group initiation, and group-target combination and splitting detection are the two special procedures in group-target tracking. They are equivalent to the pretreatment of measurements. The splitting group detection and group initiation implements the initiation of group-target tracks by correlating observation data. The combination and splitting detection is used for the association of observation data and tracks or tracks and tracks and implements track combination or splitting. In short, the splitting group detection and group initiation adopts the association of observation data and observation data; the combination of new members adopts the association of observation data and tracks; the combination of group-targets with stable tracks adopts the track–track association. The group-target splitting adopts presetting sub-wave gates within the association gates of group-targets to conduct double association.

Based on the combination and splitting detection of traditional formation targets, this chapter studies the factors which influence the combination and splitting of group-targets; proposes an independent double-threshold combination algorithm according to direction and hidden velocity information (namely a three-threshold combination algorithm), a splitting detection algorithm by presetting sub-wave gates within the association gates of group-targets to conduct double association, and a stable transition algorithm based on centroid correction according to equivalent measurements; and proposes a group termination algorithm.

6.3   Analysis of Formation Target Combination and Splitting Detection Method

6.3.1   *Formation Target Combination and Splitting Detection Method Based on Marginal Tracks*

References [8, 9] discuss the tracking of formation targets, which actually track that center and margin of the formation targets. Three tracks need to be built up for formation tracking. The first is the center track of the formation target, and the second is the two marginal tracks, as shown in Fig. 6.1. When the formation tracking starts, we should determine which single targets can constitute the formation target and separate them from the environment. After the formation target is obtained, the mean location of all targets needs to be computed, the center measurement needs to be obtained by highlighting the center, and the location of the marginal targets of the formation needs to be computed. Then, formation tracking is achieved by constructing three tracks according to the measurements of the center and two marginal targets.

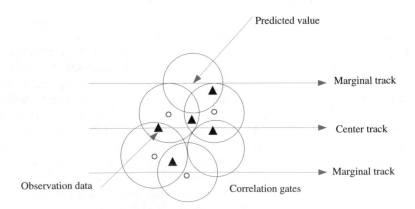

Fig. 6.1 Formation target tracking demonstration

The combination and splitting detection of formation target members is performed according to the number of the targets within the tracking beams of the marginal targets. When a few targets occur within the marginal beam, the targets that are close to the formation center need to be eliminated and those that are far from the center need to be kept, and thus, the size of the formation target increases. When the marginal measurements disappear for a few cycles, marginal targets split. In the process of tracking, whenever missing detection and/or false alarms (need to be validated for several times, in order to determine whether missing tracking is caused by missing detection or false alarms due to disturbance or noises) appear, unidentified targets become identifiable, or a few targets become unidentifiable, the formation scale all needs to be adapted in time.

The main limitation of the method is that it can only detect the splitting of marginal targets, but fails in the splitting of targets in other directions. Obviously, the methods cannot meet the requirement of group-target combination and splitting detection.

6.3.2   Formation Target Combination and Splitting Detection Method Based on Target Distance

The new-and-old member combination and splitting method based on the association of observation data and tracks of formation targets correlates the measurements obtained in every radar cycle with each formation member through highlighting the formation center by radar beams. If the method succeeds in association under the restrictions of given threshold, we keep the measurements. If the method fails, we need to process the measurement otherwise.

Assume that radar receives M valid returns (in measurement coordinate system), and we have the following:

$$Y_i(k) = (R_i, \alpha_i, \beta_i) \quad i = 1, 2, \ldots, M \tag{6.1}$$

There were N members in the formation. Now the M measurements $Y_i(k)$ need to be correlated with N members, denoted by the following:

$$\delta(i,j) = Y_i(k) - H(k)\hat{X}_j(k/k-1) \quad j = 1, 2, \ldots, N \tag{6.2}$$

If $\delta(i,j) \leq X_G$, the return i is relevant to member j. In Eq. 6.2, $H(k)$ is the measurement matrix, and $\hat{X}_j(k/k-1)$ is the estimated value of the j th member; X_G is the threshold, determined by the measurement errors and prediction errors. If there are multiple returns fall into the threshold X_G, we adopt the nearest-neighbor method. This procedure lasts until all M measurements have been correlated with

N formation members. The association can generate four results. (1) The return i is relevant to the member j. (2) None of the return i is relevant to the member j, probably because the return is too weak of the member j or the member has split from the formation. (3) The return i is irrelevant to any member j and is called redundant return. (4) The return i is relevant to the redundant member j, and this is the determination in the validation of this radar scan. The action of radar scan can determine whether redundant returns can be regarded as the action of a formation member, which is called the validation.

If each measurement is relevant to one of the formation members, the splitting detection is performed. The distance splitting threshold is set to be R_0 (is up to the designed parameters and false-alarm ratio of signal processors), and then

$$|R_c(k) - R_i(k+1)| > R_0 \qquad (6.3)$$

The splitting occurs when Eq. 6.3 holds. $R_c(k)$ is the distance measurement of the group-target centroid, $R_i(k+1)$ is the distance measurement of the return i. The angle-splitting detection is performed after obtaining the measurement of centroid return, and the determinant principle is as follows:

$$\sqrt{[\Delta \sin \alpha_i(k)]^2 + [\Delta \sin \beta_i(k)]^2} \geq \Delta_0 \qquad (6.4)$$

where

$$\Delta \sin \alpha_i(k) = \sin \alpha_c - \sin \alpha_i(k+1) \qquad (6.5)$$

$$\Delta \sin \beta_i(k) = \sin \beta_c(k) - \sin \beta_i(k+1) \qquad (6.6)$$

$\sin \alpha_c(k)$ and $\sin \beta_c(k)$ are the direction and pitching angle values at time point k. $\sin \alpha_i(k+1)$ and $\sin \beta_i(k+1)$ are the direction and pitching angle measurements of the ith return of the group-target. Δ_0 is the angle-splitting threshold.

In fact, the formation target centroid mentioned in Ref. [2] is the geometry center of the formation target. There is no filtering of candidate returns within the association gates. And every target track needs to be known. Therefore, the principle of a formation target membership is to determine between every two measurements according to the restriction of the prescribed distance threshold [2]. This combination and splitting detection method asks for many associations and awareness of the track of each target. According to the definition of the group-target, the order tracking of a formation target cannot endure the requirement of group-target tracking, and the combination and splitting detection method cannot be directly applied to the combination and splitting detection of group-target tracking. Therefore, new combination and splitting detection algorithm needs to be studied.

6.4   Group-Target Combination Algorithm

6.4.1   Observation Data and Track Combination Algorithm Based on Association Gate Principles

As we know, the observation data that appear in the last cycle and completely belong to each group-target are regarded as an existing member, and one which newly emerges in the last cycle and is correlated with none of the group-target tracks is regarded as a redundant member [2]. Similarly, this book takes newly emerging observation which correlates with some group-target tracks as a new member of the group-target. As for redundant members, group-splitting detection and group initiation is needed to determine whether they make new group-targets, and new group-target tracks are initiated. This issue was discussed in Chap. 2 in detail. The next focus is the ascription of the measurements of exiting members and new members, namely the combination detection of group-targets.

Assume that M valid returns, $Y_i(k)$ $(i = 1, 2, \ldots, M)$, have been received in a scan of radar at time k. And N group-targets come into being after measurements pretreatment. Now we need compare every $Y_i(k)$ of the M measurements with the estimated centers of the N group-targets. The operation is denoted by

$$g_{ij}(k) = Y_i(k) - H(k)\hat{X}_j(k/k - 1) \leq \gamma_j \quad (i = 1, 2, \ldots, M; \; j = 1, 2, \ldots, N) \quad (6.7)$$

If $g_{ij}(k) \leq \gamma_j$, the measurement is correlated with the group-target j. $H(k)$ is the measurement matrix. $\hat{X}_j(k/k - 1)$ is the forecast value of the equivalent measurement of group-target j. γ_j is the association threshold of group-target j. This operation continues until the end of all comparisons between each measurement and each forecast center of the group-targets.

The association can generate three results. (1) If the measurement number received in this cycle is greater than that was obtained in the last cycle, the newly emerging returns can be new targets, targets which become identifiable at this time while cannot be identified before, or even false measurements. (2) If the measurement number received in this cycle is less than that was obtained in the last cycle, targets might be unidentifiable due to too little distance, shelter might occur between targets, or missing detection occurs. (3) If the measurement number received in this cycle is equal to that was obtained in the last cycle, either the simplest situation that targets and returns in the neighboring cycles are correlated with each other one by one occurs or targets are unidentifiable (due to too little distance)/sheltered (by other targets) to decrease the actual measurements. The number of false measurements is accidentally equal to the number of decreased measurements, which occurs with very small probability. So this small probability event is neglected in this book.

The dealing principle to the abovementioned situations is as follows. (1) All the returns are correlated with the corresponding group-targets, so that we compute the equivalent measurement of all measurements to be used by the filter updating in the next cycle. (2) The equivalent measurement of all the returns that are successfully correlated with part of the group-targets is computed. The group-targets without return association directly use the equivalent measurement of the last cycle to update filters. The remaining returns that belong to no group-targets will be validated in the next cycle. If they are still uncorrelated, they will be treated as new targets and their tracks will be initiated. (3) As for those that are correlated with multi-group-target tracks, the measurements will be processed according to the common-measurement principles for multi-group-targets discussed in Chap. 4.

6.4.2 Track–Track Combination Algorithm Based on Direction and Hidden Velocity Restrictions

According to the definition of a group-target, we set the combination condition. When the distance of two group-targets is less than the maximum of the thresholds of the association gates of the two when association gates fork, namely one centroid of the two falls inside the association gate of the other, the two group-targets can be combined. We call the principle of one-by-one combination the principle of group-target track combination. The combination of tracks that is similar to the combination of observation data and tracks is an important aspect of group-target combination. The two combinations can exist independently or concurrently.

As for the combination of group-targets, traditional methods depend only on whether the distance between two closest observation data among the members of the two group-targets is less than predefined target distance. This method obviously can deal with only the situation when the two group-targets are completely parallel. However, the combination is unnecessary under a similar location-satisfied situation, when the tracks of the two group-targets are forking or in contrary directions. The combination of group-targets essentially represents the location-displacement feature and non-identifiability of the two group-targets within a given time period. Since the formation target combination method considers no motion direction and velocity difference of group-target tracks, the method can easily lead to mistaken combination when applied to group-target combination. Consequently, we need to judge whether group-target should be correctly combined in comprehensive consideration of the directions, velocities, history, and location information of tracks. Although group-target tracking is implemented based on single sensors, the combination of group-targets essentially represents the combination of tracks, and additionally, the equivalent measurement of the newly emerged group-targets should depict the mean mobility feature of all members. Therefore, we apply the algorithms that describe the track association from distributed multi-sensors to the group-target combination based on a single sensor.

As we know, in the process of distributed multi-sensors information fusion, there are multiple algorithms relevant to track association, such as weighting method, sequential track association method, nearest-neighbor and K-neighbor track association algorithm, distribution method, statistical track association algorithm on multiple local nodes, relevant double-threshold track association algorithm, and independent double-threshold algorithm. Among them, the independent double-threshold algorithm has the highest correct association probability except for a bit heavier computational burden [125, 126]. The double-threshold detection is derived from the signal detection in radar automatic detection theory. It compares the range of some returns with the first threshold one by one; accumulates the counter when a sample exceeds the threshold; compares the result of the counter to the specified second threshold; and determines the existence of the signal when the result is greater than the second threshold, otherwise abandons the signal.

Based on the double-threshold association method, we propose a three-threshold combination algorithm in track–track combination, considering the track direction information of two group-targets. Although the algorithm involves only track direction but velocity information, it already implicitly involves velocity restriction. Then, we will explain in detail.

Considering any two group-targets l, h, assume that: (1) The first threshold is θ_0 and ϑ_0, respectively; (2) the second threshold is the association threshold of the corresponding group-targets, denoted by γ_l or γ_h; (3) the validation times is F, and the three threshold is ΔF; (4) the direction intersection angle of the group-target l, h is $\Delta\theta(k), \Delta\vartheta(k)$; and (5) the measurements of the two group-targets have been calibrated to time. The equivalent measurement state and prediction of group-targets are denoted by the following:

$$
\begin{cases}
\widehat{X}(k) = \left[\hat{x}(k)\hat{y}(k)\hat{z}(k)\dot{\hat{x}}(k)\dot{\hat{y}}(k)\dot{\hat{z}}(k)\right] \\
\widehat{X}(k/k-1) = \left[\hat{x}(k/k-1)\hat{y}(k/k-1)\hat{z}(k/k-1)\dot{\hat{x}}(k/k-1)\dot{\hat{y}}(k/k-1)\dot{\hat{z}}(k/k-1)\right] \\
Y(k) = [x(k)y(k)z(k)]
\end{cases}
$$

$$(6.8)$$

First, perform intersection angle judgment, namely the first threshold detection, before the combination detection of group-target tracks. According to Eq. 6.8, we get the following:

$$
\begin{cases}
\Delta\theta(k) = \arctan\left(\dfrac{\hat{\dot{y}}_l(k/k-1)-\hat{\dot{y}}_h(k/k-1)}{\hat{\dot{x}}_l(k/k-1)-\hat{\dot{x}}_h(k/k-1)}\right) \\
\Delta\vartheta(k) = \arctan\left(\dfrac{\hat{\dot{z}}_l(k/k-1)-\hat{\dot{z}}_h(k/k-1)}{\sqrt{(\hat{\dot{x}}_l(k/k-1)-\hat{\dot{x}}_h(k/k-1))^2 + (\hat{\dot{y}}_l(k/k-1)-\hat{\dot{y}}_h(k/k-1))^2}}\right)
\end{cases}
$$

$$(6.9)$$

Then, if the two group-targets satisfy $\Delta\theta(k) \leq \theta_0$ and $\Delta\vartheta(k) \leq \vartheta_0$, we continue detection, otherwise stop here. Since the two group-targets' equivalent measurements are independent and under the same coordinate system, the statistical distance after time adjustment is as follows:

$$\Delta X(f) = \left[\hat{X}_l(f) - \widehat{X}_h(f)\right]^T [P_l(f) + P_h(f)]^{-1} \left[\hat{X}_l(f) - \widehat{X}_h(f)\right] \quad (f = 1, 2, \ldots, F)$$
(6.10)

If:

$$\Delta X(f) \leq \gamma_l \text{或} \gamma_h$$
(6.11)

We get:

$$\Delta F_{lh}(f) = \Delta F_{lh}(f - 1) + 1, \quad \Delta F_{lh}(0) = 0$$
(6.12)

When $\Delta F_{lh}(f) \geq \Delta F$, the two group-targets l, h are possibly combined to one group-target; otherwise, they are processed as unlikely to be combined. $\hat{X}_l(f), \widehat{X}_h(f)$ and $P_l(f), P_h(f)$ in Eq. 6.10 are called the state approximation and the covariance of approximation error of the group-targets l, h. $\Delta F/F$ is generally set to be 6/8 [76].

Please note that the independent double-threshold detection incorporates the second and third threshold detection, implicating velocity restrictions. Forking indicates that the two tracks definitely have direction and/or velocity differences. Since the direction factor has been taken as the first threshold to constrain, only velocity difference need to be considered here. According to the rule given in Ref. [2], the velocity difference of two tracks should obey the velocity principle, namely that they cannot exceed the ratio between the group-target association gate and the radar cycle. If the velocity difference is beyond the velocity threshold, the comparative locations of the two tracks in one cycle will dissatisfy the second threshold. Therefore, the combination detection implicates velocity restriction so that we unnecessarily need to consider the influence of velocity again.

If $\Delta F_{lh}(f)$, the times when the association of the group-targets l, h satisfies the second threshold at time f, is called the association quality of the group-targets, we get the splitting quality of the group-targets l, h:

$$\Delta E_{lh}(f) = \Delta E_{lh}(f - 1) + 1$$
(6.13)

The splitting quality indicates the times when the association of the group-targets l, h dissatisfies the second threshold at time f. Adopting association quality and splitting quality can help real-time control of the process of the association detection. If at time $f - 1$ we get the following:

$$\Delta E_{lh}(f - 1) > F - \Delta F$$
(6.14)

Then, association detection stops at the next time f. Whatever the detection result is after time $f - 1$, the association result of the group-targets l, h definitely dissatisfies the third threshold until the number of detection times completes. In other words, there must exist $\Delta F_{lh}(F) < \Delta F$ until the completeness of F times of detections. Likewise, if at time $f - 1$ we get the following:

$$\Delta F_{lh}(f - 1) \geq \Delta F \tag{6.15}$$

Whatever the detection result is after time $f - 1$, the association result of the group-targets l, h definitely satisfies the third threshold until the completeness of F times of detections. When two or more group-targets satisfy the first and the second thresholds, ambiguity can occur in the association detection, so we need keep performing association detections until the completeness of F times of detections. If multi-group-targets all satisfy the third threshold, we combine all of them; otherwise, we combine only those that satisfy the third threshold. If Eqs. 6.14 and 6.15 are not satisfied, we still need continue until the completeness of F times of detections.

Please note the difference between group-target combination and observation data aggregation. According to Refs. [106–108], radar signals usually need to go through pulse pressure, noise wave elimination, accumulative detection of slipping windows and videos, horizontal false-alarm detection, and mobile-target detection, to restrain noise waves and distill signals. In fact, even we comprehensively adopt several processing methods, strong noise wave, objects on the floor, and weak noise waves from similar targets will unexceptionally leave some amount of residual observation data, which brings difficulties in dealing with radar data especially for close distance targets and big targets. Since radar can give no exclusive video envelope, multi-returns appear and consequently influence the resolution of distance and angle and measure precision of radar. Therefore, the observation data aggregation techniques that can improve radar resolution and measure precision occur. The observation data aggregation includes 4 steps: (1) distilling of Doppler information and target property information, (2) correlating in distance and generating group reports, (3) correlating group reports in angles and generating observation data reports, and (4) adding information confidence to the observation data reports and deleting some assured noise waves, which is also called post-filtering process. The essence of the observation data aggregation technique is to process single targets at the data layer, for the sake of improving radar resolution and tracking precision. Although group-target combination processes each member of the group-target at the data layer, this process is to obtain the centroid (equivalent measurements) of the group-target according to the all-neighbor idea of the PDA algorithm, utilizing the distance between each target measurement and the predicted center of the association gate of the group-target under the restriction of association gates. Therefore, although the algorithm of observation data aggregation and the algorithm of group-target centroid are very similar, their concerned numbers of targets and aims are different. The essence of group-target combination is to deal with multi-targets in high density at the data layer, in order to assure the surveillance and tracking of all targets.

6.5   Algorithm of Group-Target Splitting Detection

6.5.1   Main Idea of Group-Target Splitting Detection

For remote distance surveillance system, predicting as early as possible the splitting situation of group-target members under surveillance and tracking is critical to obtaining more prewarning time and correct judgment of enemy intention for the defense side. Traditional formation target splitting detection methods track formation targets' marginal measurement because targets located at the margins of the formation are supposed to split [2, 8]. The methods track the center of a formation target and, meanwhile, track the marginal measurements of the formation target by two beams. Whether formation members split is determined according to specified target distance principles. Thus, these methods can only detect the splitting situation of the targets on the two sides of a formation target. However, when a formation target is located at the heading or tailing of radar, these methods can only detect the splitting on two sides of the formation instead of the front, upper side, and lower side of the formation. In particular, they cannot detect the splitting situation on the front and lower side of the formation. When a formation is close to the tracking convenient location of radar coordinates, only the front members of the formation can be detected to split, because the formation target is almost unlikely to withdraw from the back. As we know, the changes in the location of group-targets' marginal targets will directly influence the scale of the group-targets. This somehow reflects the maneuver of part of the marginal members, and this maneuver can be regarded as the sign of splitting of group-target members. As a matter of fact, according to tactical demand and the relative locations between radar and group-targets, the splitting of group-targets' marginal members will not appear in all directions, and inner members can likely to start splitting (e.g., the concurrent shooting of a plane formation); namely, all the members of a group-target can split in all directions.

The essence of group-target member splitting indicates the maneuver of part of the members, which means the changes in locations, velocities, and directions. Namely, splitting targets can have changes in the radial and/or centripetal accelerations. The velocity information represents the radial movement of targets, while the direction information represents the centripetal movement of targets. The location, velocity, and direction constitute the three elements of the splitting detection of group-target members. In order to improve the traditional methods in the inadaptability of all-direction splitting detection of members, we need to get the location, velocity, and direction in the first place.

We analyze the three elements of location, velocity, and direction as follows. (1) It is easy to get location information so we neglect here. (2) It is easy to obtain the holistic velocity information of a group-target through difference in neighboring cycles, but it is uneasy to obtain the velocity information of each measurement. (3) Direction information somehow represents the centripetal movement of targets, which is most difficult to obtain. Although some references [47–55] propose to map the measured spatial data to those in the pattern space through wavelet transform for the sake of obtaining direction information, it is computationally intractable. However, the direction information collection can be achieved considering the fake direction conceptions and the location-displacement feature of group-targets discussed in references [125, 126].

Based on the idea, we draw a line or other figures, among the group-target members that fall into the association gates in the neighboring cycles, to describe the relationships among members to get direction information. To do so, we make the following attempts.

First, compare the geometric figure incorporating the lines drawn between the marginal measurements in the same cycle, with those drawn in the next cycle with the same means. We can detect whether splitting occurs by judging the degree of the resemblance between the figures. This method is intuitionistic, but it is affected significantly by the false measurements and cannot describe the splitting situation of non-marginal members, as shown in Fig. 6.2.

Second, compare the parallelism between the fake directions in the last cycle and this cycle. This method is simple, but it needs not only time of three cycles and the maneuver detection of group-targets which complicates the splitting detection, as shown in Fig. 6.3.

Third, compare the parallelism between the lines drawn between the measurements in neighboring cycles. This method cannot determine which measurement in the last cycle is correspondent to which measurement in this cycle, and each measurement constituting the lines can be noise wave or false alarm. Therefore, this method is infeasible and inapplicable, as shown in Fig. 6.4.

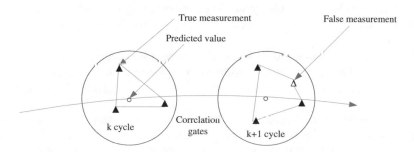

Fig. 6.2 Demonstration of the resemblance judgment in neighboring cycles

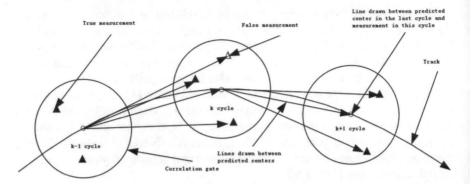

Fig. 6.3 Demonstration of the fake direction parallelism judgment in neighboring cycles

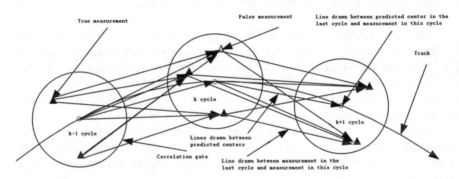

Fig. 6.4 Demonstration of the parallelism judgment of the lines drawn between measurements in neighboring cycles

Fourth, map the measured spatial data to that in the pattern space through wavelet transform [46–53], obtain direction information in the pattern space, and transform the information again to the measurement space through wavelet backward transform. However, this method is computationally intractable and cannot be applied well to the direction information distilling of group-target splitting members.

Fifth, the group-target imaging detection method is computationally intractable and insufficient in the precision of splitting detection (here is not specially focused on high-resolution radar). Therefore, this method is inapplicable in the splitting detection.

All the methods abovementioned have shortcomings and can dissatisfy the requirement of obtaining direction information, through comparison and analyses. Further studies find out that stable group-target members and equivalent measurements have common location-displacement features, while members that have maneuver (are likely to split) have bad location-displacement features. So we draw lines between the predicted center of equivalent measurements in the same cycles

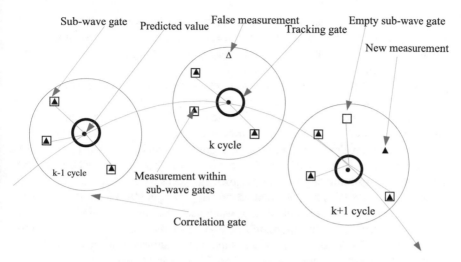

Fig. 6.5 Demonstration of the splitting detection based on sub-wave gates

and every measurement, make equal parallels in the next cycles, take the ends of the parallels as predicted values, construct sub-wave gates by using some correct receiving probability as thresholds, and achieve the occurrence detection of maneuver. In the meantime, we perform a second filtering by constructing sub-wave gates within association gates, obtain the targets with inconsistent location-displacement features, and possess the power of restraining false measurement with corresponding disposal. We investigate this method in depth as shown in Fig. 6.5.

Adopting the abovementioned idea, we propose an idea of presetting sub-wave gates within the association gates to conduct double association splitting detection. (1) Draw lines between the centers of association gates in $k-1$ cycle and all measurements within association gates in this cycle. (2) Taking the predicted centers of equivalent measurements in k cycle as ends, draw segments that are parallel to and equal in length to the lines drawn between the ends and all measurements in the last cycle. (3) Taking the other end as the predicted center of the measurements in the last cycle, determine the threshold considering the direction information based on fake directions and velocity information based on mean displacement (to assure some correct receiving probabilities), and construct sub-wave gates of direction velocity. (4) The measurements fallen into sub-wave gates in k cycle are regarded to have good location-displacement features, while those fallen outside sub-wave gates are possibly false measurements, new measurements, or splitting measurements. We give the latter some "penalty," by assigning a penalty factor (less than 1) according to its distance to the predicted center of a group target. (5) Adjust the weights (in the meantime normalizing the weights again) of the measurements outside sub-wave gates according to the penalty factor, and set the equation to compute the equivalent measurements of group-targets under the situation of members' splitting. The method utilizing the penalty factor to correct has the merit of real-time watching of the moving state of each target, which avoids the unlimited enlargement of association gates, favors the quick splitting of members from group-targets, and grips the moving state of inner members of group-targets (furthermore, the synthesized track of all targets within each sub-wave gate can be given). The length of the line drawn between the predicted value in $k - 1$ cycle and the measurement in k cycle is the so-called fake velocity (based on this, the comparatively precise velocity of each target can be reasoned considering geometric relationships). And we also know the fake direction and velocity of splitting targets, so that quick and precise track initiation of splitting targets can be carried out. The method utilizing the penalty factor to correct the weights of the measurement outside sub-wave gates can also decrease the influence of the group-targets' splitting members and false measurements on equivalent measurements. The introduction of sub-wave gates multiplies all members outside sub-wave gates by a coefficient (less than 1) no matter they are false, new, or splitting measurements. This equally degrades the contribution of these measurements to equivalent measurements. Therefore, this method not only performs step-by-step approaching to members that split from group-targets and assures continuous "penalty" to splitting members, but also guarantees them to split from group-targets as soon as possible (this is consistent with the group-target tracking principles, making single-target tracking as early as possible). This is good for the stable transition (avoiding great fluctuations) of group scale when guaranteeing the splitting of members from group-targets. Additionally, the introduction of the penalty factor to the measurements outside sub-wave gates equally restrains false measurements such as noise waves and degrades the influence of these false measurements on equivalent measurements.

Furthermore, the feature of centripetal maneuver is having direction changes but almost no radial changes. So the introduction of direction information correspondingly gives the information of centripetal movement of targets and thus solves the difficulty of identifying whether targets have maneuver splitting due to Doppler frequency displacements (distance changing rate is zero when targets have cyclic movements). Likewise, the feature of radial maneuver is having almost no direction changes, but radial distance changes. So the introduction of velocity information improves the capability of radially detecting targets' splitting. When targets have both centripetal and radial maneuver, direction and distance both changes, so the sub-wave gates of direction velocity can completely deal with the target splitting situation and improve the capabilities of target splitting detection and noise wave restrains.

This method of presetting sub-wave gates inside association gates equally obtains approximately the tracks of the members of group-targets before splitting. The sub-wave gates incorporating fake velocity and direction information can determine the approximate location of each member before it splits. The line drawn between these locations is naturally the targets' track within the sub-wave gate.

Clearly obtaining the approximate synthesized tracks of a single target or multiple close targets within the sub-wave gates is crucial to get the splitting state of group-targets' members as early as possible and makes great contribution.

Therefore, the three elements of group-target splitting detection can be obtained, and we will investigate the splitting detection algorithms.

6.5.2 Double Association Splitting Detection Algorithm by Presetting Sub-Wave Gates Within the Association Gates

1. Construction of direction—velocity sub-wave gates

Direction and velocity are used to determine whether the same measurement within association gates in neighboring cycles belongs to the same target. Direction gives the centripetal maneuver information of targets, while velocity gives the radial maneuver information of targets. The sub-wave gates construct by direction and velocity can be used to determine the splitting situation of targets between measurements in neighboring cycles. The construction of sub-wave gates within association gates is as follows. (1) Draw a parallel which is parallel to the line between predicted center in $k - 1$ cycle and measurements within its association gates, and is equal to the direction segment (the line between the predicted center and measurements). (2) Within the association gates in k cycle, centered on the end of the direction segment, the probability of falling into sub-wave gates with certain direction and velocity is set to be the threshold, which leads to the equality of the number of the measurements in k cycle to the number of sub-wave gates within association gates.

The processing of the measurements inside and outside sub-wave gates is described next. (1) The measurement fallen inside sub-wave gates is correlated with that in $k - 1$ cycle, and all the measurements fallen inside association gates and outside sub-wave gates can be false, new, or sheltered measurements. (2) As for the situation when multiple measurements fall inside the same sub-wave gate, considering the multi–multi correspondence between targets and measurements in this book, the observation data that are closest to the predicted center of sub-wave gates are taken as the true measurement, and the rest are taken as the residual returns of targets. This is called the residual measurement of multi–multi correspondence (please note that the number of all measurements is considered when computing equivalent measurements). (3) All measurements outside sub-wave gates are assigned with a penalty factor based on direction and velocity which is less than 1. Then, the results of splitting detection based on sub-wave gates can be the following.

① All measurements in $k-1$ cycle are corresponding one by one to those in k cycle, and all measurements fall inside sub-wave gates. This indicates the normal moving state of group-targets and no splitting.

② The number of all measurements in $k-1$ cycle, Γ_{k-1}, is equal to Γ_k in k cycle. But empty sub-wave gates appear which indicates possible splitting.

③ The number of all measurements in $k-1$ cycle, Γ_{k-1}, is greater than Γ_k in k cycle. There are empty sub-wave gates and no returns outside sub-wave gates, which indicates missing alarms or target shelters.

④ The number of all measurements in $k-1$ cycle, Γ_{k-1}, is less than Γ_k in k cycle. Sub-wave gates are not empty, and there are returns outside sub-wave gates, which indicates false measurements, reoccurrence of sheltered targets, or new targets.

⑤ The number of all measurements in $k-1$ cycle, Γ_{k-1}, is less than Γ_k in k cycle. Sub-wave gates are empty, and there are returns outside sub-wave gates, which indicates target splitting and false measurements, reoccurrence of sheltered targets, or new targets.

Making abovementioned judgments one by one is very difficult. The true measurements fallen outside sub-wave gates deviate in direction, velocity, or both, or they are false measurement. The commonness is that the true measurements all fallen outside sub-wave gates. Therefore, we propose the equivalent measurement algorithm considering penalty factor and weight adjustment. All measurements fallen outside sub-wave gates are multiplied by a penalty factor considering direction and velocity, namely, to weaken the members with mobility characteristic differences of group-targets or false measurements. This is also can be seen as the parallel displacement result of the measurements fallen inside sub-wave gates, while these measurement have the consistent feature to the mean mobility of group-targets. Otherwise, the measurements have different mobility features, and splitting and/or false measurement can exist. This method equivalently performs double return filtering based on sub-wave gates.

First give the computational equation of predicted center of sub-wave gates, according to the projection theory of right-angle coordinate system.

$$Y_i^c(k) = Y^e(k) - Y^e(k-1) + Y_i(k-1) \qquad (6.16)$$

where $Y_i(k-1)$ and $Y^e(k-1)$ are the predicted values of ith measurement and equivalent measurement in $k-1$ cycle, respectively. $Y^e(k)$ and $Y_i^c(k)$ are the predicted centers of equivalent measurements of group-targets in k cycle and the ith measurement in $k-1$ cycle, respectively. The size of sub-wave gates will be discussed after obtaining the predicted center. According to the location-displacement feature of group-target members, every true measurement will be scattered inside the sub-wave gates with its own dispersion. Therefore, draw a line between $Y^e(k)$—the predicted center of equivalent measurements of group-targets in $k-1$ cycle and

$Y_i^c(k)$—the predicted center of the ith measurement in k cycle, and a line between $Y^e(k)$—the predicted center of equivalent measurements of group-targets in $k-1$ cycle and $Y_i(k)$—the measurement in k cycle, where $Y_i(k)$ has three components $x_i(k), y_i(k), z_i(k)$. The intersection angle between the two lines is called direction difference, denoted by the following:

$$\begin{cases} \delta_{i\theta}(k) = \arctan\left(\frac{y_i^c(k)-y^e(k-1/k-2)}{x_i^c(k)-x^e(k-1/k-2)}\right) - \arctan\left(\frac{y_i(k)-y^e(k-1/k-2)}{x_i(k)-x^e(k-1/k-2)}\right) \\ \delta_{i\vartheta}(k) = \arctan\left(\frac{z_i^c(k)-z^e(k-1/k-1)}{\sqrt{[x_i^c(k)-x^e(k-1/k-2)]^2 + [y_i^c(k)-y^e(k-1/k-)]^2}}\right) - \\ \arctan\left(\frac{z_i(k-1)-z^e(k-2/k-2)}{\sqrt{[x_i(k)-x^e(k-1/k-2)]^2 + [y_i(k)-y^e(k-1/k-2)]^2}}\right) \end{cases} \quad (6.17)$$

The measurements inside sub-wave gates are assumed to be true measurement, so they obey Gauss distribution. Equations 3.19–3.22 in Chap. 3 discussed the standard error of true measurement. To transform them into a direction angle, when Eq. 6.17 is satisfied:

$$\begin{cases} \delta_{i\theta} = y_i(k) - y_i(k-1)/r_i(k) < K_G\sigma_o/r_i(k) \\ \delta_{i\vartheta} = z_i(k) - z_i(k-1)/r_i(k) < K_G\sigma_o/r_i(k) \end{cases} \quad (6.18)$$

We determine that there is no centripetal maneuver for group-targets at time k, compared to the situation at time $k-1$. $r_i(k)$ is the distance of ith measurement.

Since group-target tracking can only give the radial velocity of equivalent measurement, it cannot get the velocity of each measurement. Namely, it can only get the parallel-displacing velocity of two parallels. So the velocity gate is computed by dividing the direction-moving distance by radar cycle. According to Fig. 6.5 and Eq. 6.16, the velocity threshold is as follows:

$$\delta_{iV} = |Y^e(k/k-1) - Y^e(k-1/k-2)|T^{-1} < K_G\sigma_o T^{-1} \quad (6.19)$$

In Eq. 6.18,

$$r_i(k) = |Y_i(k) - Y^e(k/k-1)| \quad (6.20)$$

We finally get the dimension of sub-wave gate:

$$G_i = \begin{cases} \delta_{i\theta} < K_G\sigma_o/r_i(k) \\ \delta_{i\vartheta} < K_G\sigma_o/r_i(k) \\ \delta_{iV} < K_G\sigma_o/T \end{cases} \quad (6.21)$$

Equation 6.21 can be also called the threshold when no splitting occurs from time $k-1$ to time k, and all the measurements within the threshold are regarded as true measurements. Please note that some references adopt the line between the state approximation in the last cycle and observation data in this cycle, which can lead to the difference between the initiation spots in the predicted direction. Here, we use the fake direction obtained from the line between the predicted center in the last cycle and the observation data in this cycle. Note that K_G should be less than K_t; otherwise, sub-wave gates can be bigger than association gates.

2. Construction of association matrix based on direction velocity sub-wave gates

Construction of association matrix of direction velocity is to determine which measurements within association gates fall inside sub-wave gates and which fall outside. Equations 6.16 and 6.21 give the predicted center and threshold of sub-wave gates. Based on the observation data obtained at time $k - 1$ to form the sub-wave gate at time k, compare it with the measurements obtained at time k one by one, and we get "1" the threshold value is satisfied, otherwise get "0." By assuming N sub-wave gates $Y_i^c(k)$ come into being at time k according to the situation at time $k - 1$, and m_k measurements $Y_j(k)$ are received at time k where $i = 1, 2, \ldots, N$ and $j = 1, 2, \ldots, m_k$. Then, the direction velocity association matrix is as follows:

$$
\begin{bmatrix}
 & 1 & 2 & \cdots & m_k \\
1 & \Delta_{11}(k) & \Delta_{12}(k) & \cdots & \Delta_{1m_k}(k) \\
2 & \Delta_{21}(k) & \Delta_{22}(k) & \cdots & \Delta_{2m_k}(k) \\
\vdots & \vdots & \vdots & \cdots & \vdots \\
N & \Delta_{N1}(k) & \Delta_{N2}(k) & \cdots & \Delta_{Nm_k}(k)
\end{bmatrix}
\tag{6.22}
$$

The value of each element in Eq. 6.22 is as follows:

$$
\Delta_{ij}(k) = \begin{cases} 1(|Y_j(k) - Y_i^c(k)| < G_i) \\ 0(|Y_j(k) - Y_i^c(k)| \geq G_i) \end{cases}
\tag{6.23}
$$

Add up the elements in one column in Eq. 6.22, and compute the norm of the first column to get the number of sub-wave gates that some measurement falls inside the following:

$$
\nabla_j(k) = \sum_{i=1}^{N} \Delta_{ij}(k)
\tag{6.24}
$$

Add up the elements in one row in Eq. 6.22, and compute the norm of the first row to get the number measurements that fall inside each sub-wave gate:

$$
\nabla_i(k) = \sum_{j=1}^{m_k} \Delta_{ij}(k)
\tag{6.25}
$$

Add up the elements in all rows and columns to get the number of measurements that fall inside sub-wave gates:

$$\nabla_{ij}(k) = \sum_{i=1}^{N} \sum_{j=1}^{m_k} \Delta_{ij}(k) \tag{6.26}$$

The measurements that correspond to the sum of all the elements in jth column, $\nabla_j(k) = 0$, are the measurements outside all sub-wave gates and will be processed with their weights.

Since one of the important hypotheses is the multi–multi corresponding between targets and measurements, the principle of determining the number of sub-wave gates in the next cycle is as follows. (1) Multiple measurements can fall inside a sub-wave gate, but the predicted center will be given in the next cycle considering only the measurement which is closest to the predicted center of the sub-wave gate. (2) When sub-wave gates fork, the belonging of the measurements falling inside forked sub-wave gates will be determined according to the nearest-neighboring principle. And the predicted center will be given in the next cycle considering only the measurement which is closest to the predicted center of the forked sub-wave gates. (3) When only one measurement falls inside the intersected area of two sub-waves, it will be determined to belong to the nearest sub-wave gate according to the nearest-neighboring principle. Please note the nearest-neighboring principle indicates that one will be selected, when its velocity is closest to that of the predicted center of sub-wave gates inside the direction velocity sub-wave.

3. Equivalent measurement algorithm based on the penalty factor of group-target splitting members

A group-target will remain unstable when the targets inside geometric margins are splitting. The gradual transition method needs to be adopted to decrease the contribution of splitting members on equivalent measurement and make the shift of the equivalent measurement location unchanged or changed a bit. The aim was to avoid the location of the equivalent measurement to go outside the tracking gate due to location mutation. Therefore, the penalty factor and the equivalent measurement algorithm based on the penalty factor are studied.

① Computation of the penalty factor

Draw a line between the predicted center of equivalent measurement at time $k - 1$ and the members falling outside sub-wave gates at time k. The intersection angle between this line and the predicted direction from $k - 1$ time to k time is called a fake direction angle. For true measurements obeying Gauss distribution, normalization of the fake direction angles can be performed using the covariance of predicted direction. The weights of the members about to split can be set accordingly. The bigger the fake direction angle is, the smaller the weight is. This weakens the contribution of the fake direction angle on the equivalent measurement of group-targets and subsequently leads to the stable transition during splitting. Setting the penalty factor based on direction information instead of location information is because angle information can provide more effectively the maneuver of the measurements outside each sub-wave gate, but location information cannot easily provide the needed maneuver information. According to references [62, 76], the fake direction angle is as follows:

$$
\begin{cases}
\varepsilon_{\theta i}(k) = \theta(k) - \arctan\left(\frac{y_i(k) - y^e(k-1/k-2)}{x_i(k) - x^e(k-1/k-2)}\right) \\
\varepsilon_{\vartheta i}(k) = \vartheta(k) - \arctan\left(\frac{z(k) - z_g(k-1/k-1)}{\sqrt{[x_i(k) - x^e(k-1/k-2)]^2 + [y_i(k) - y^e(k-1/k-2)]^2}}\right)
\end{cases}
\tag{6.27}
$$

The predicted direction of the group-targets in Eq. 6.27 is as follows:

$$
\begin{cases}
\theta(k) = \arctan\left(\frac{\hat{y}^e(k/k-1)}{\hat{x}^e(k/k-1)}\right) \\
\vartheta(k) = \arctan\left(\frac{\hat{z}^e(k/k-1)}{\sqrt{[\hat{x}^e(k/k-1)]^2 + [\hat{y}^e(k/k-1)]^2}}\right)
\end{cases}
\tag{6.28}
$$

According to references [62, 76], the probability densities of the two fake directions are as follows:

$$
\begin{cases}
\iota_{\theta i}(k) = \exp\left[-\frac{1}{2}\varepsilon_{\theta i}^T \Sigma_\theta^{-1} \varepsilon_{\theta i}\right] \\
\iota_{\vartheta i}(k) = \exp\left[-\frac{1}{2}\varepsilon_{\vartheta i}^T \Sigma_\vartheta^{-1} \varepsilon_{\vartheta i}\right]
\end{cases}
\tag{6.29}
$$

In Eq. 6.29, $\Sigma_\theta, \Sigma_\vartheta$ are the covariances of the two track directions, respectively.

$$
\upsilon_i(k) = \exp\left[-\frac{1}{2}\delta_{iV}^T \Psi^{-1} \delta_{iV}\right]
\tag{6.30}
$$

where Ψ is the covariance of the velocity residual.

Equations 6.29 and 6.30 are suitable for real measurements. But we cannot assure all the measurements that fall outside sub-wave gates are real measurements, among which false measurements can appear. False measurements obey uniform distribution. We can adopt the method proposed in Chap. 4, dividing the direction angle by the maximum angle deviation of the predicted center of equivalent measurement within association gates in this cycle, to approximate the weights.

The group-target data association adopts the hybrid method of the nearest neighboring in location and the probabilistic data association. Direction and velocity weights, namely the penalty factor is as follows:

$$\begin{cases} \rho_{\theta i}(k) = 1 - \varepsilon_{\theta i}(k)/\omega_{\theta\,\text{max}} \\ \rho_{\vartheta i}(k) = 1 - \varepsilon_{\vartheta i}(k)/\omega_{\vartheta\,\text{max}} \\ \rho_{vi}(k) = 1 - v_i(k)/v_{\text{max}} \end{cases} \tag{6.31}$$

In Eq. 6.31, $\omega_{\theta\,\text{max}}, \omega_{\vartheta\,\text{max}}, v_{\text{max}} (v_{\text{max}} = |r(k) - r(k-1)|_{\text{max}} T^{-1})$ are the line between the predicted center and the edge of association gate in the last cycle, the maximum intersection angle compared to the predicted direction, and the maximum location displacement in one cycle, respectively.

② The equivalent measurement algorithm based on the penalty factor

According to Eq. 4.57 in Chap. 4, the nearest-neighboring weights of each measurement location within group-target association gates are compared to the predicted center:

$$\rho_i(k) = 1 - \eta_i = 1 - g_i(k)/\gamma \tag{6.32}$$

According to 6.31, the association probability of measurement correction outside sub-wave gates is as follows:

$$\lambda_i(k) = \rho_i(k)\rho_{\theta i}(k)\rho_{\vartheta i}(k)\rho_{vi}(k) \tag{6.33}$$

m_k is the number of valid measurements inside association gates, $M(M = \nabla_{ij}(k))$ is the number of the members falling outside sub-wave gates.

$$\lambda_i'(k) = \frac{\lambda_i(k)}{\sum_{i=1}^{m_k-M} \rho_j(k) + \sum_{i=1}^{M} \lambda_i(k)} \tag{6.34}$$

$$\beta_j'(k) = \frac{\rho_j(k)}{\sum_{i=1}^{m_k-M} \rho_j(k) + \sum_{i=1}^{M} \lambda_i(k)} \tag{6.35}$$

The equation of computing the equivalent measurement with correction is as follows:

$$Y_m^e(k) = \sum_{i=1}^{M} \lambda_i'(k)Y_i(k) + \sum_{i=1}^{m_k-M} \beta_j'(k)Y_j(k) \tag{6.36}$$

In short, the method of presetting sub-wave gates within the association gates to conduct double association and correcting with the penalty factor to the measurements outside sub-wave gates has several advantages. (1) Avoid unlimited enlargement of association gates. Correcting the weights of the measurements outside sub-wave gates by the penalty factor is to avoid the drift of the group-target centroid with splitting members, making the centroid move appropriately toward splitting members while keeping the stability of the centroid, and consider the targets with different movement states. This favors the quick splitting of members from group-targets (consistent with the principles in the group-target tracking algorithm) and the stable transition of group-target centroid during splitting. (2) Watch in real time the moving state of each target, including both the marginal members and the inner members of group-targets. (3) Solve the incapability of performing splitting detection when radial velocity is zero, namely the problem faced by Doppler radar. (4) Obtain the tracks of the members that do not split from group-targets under the given conditions. (5) The awareness of the location and velocity information of group-targets is useful to the quick initiation of the splitting members of group-targets. (6) Have the capability of restraining noise waves, which makes the group-target tracking algorithm is applicable not only to group-target tracking and splitting detection, but also to multi-target tracking in high density under the condition of dense multi-returns. Because of the capability of noise wave restraining due to the double association of sub-wave gates, or say processing of false-alarm and missing detection in multi-target tracking, can complement to some extent the incapability of distinguishing real measurements from the false measurements. And this is consistent completely with the all functions of traditional multi-target tracking algorithm.

Please note that group-target splitting detection is different from the identification of low-resolution radar based on formation target number at the signal layer. Group-target splitting detection takes advantage of the location, velocity, and direction information of each member, implementing target splitting at the data layer. Target identification of low-resolution radar takes advantage of Doppler Beam Sharpening [178] and Wavelet Transform [179], implementing target splitting at the signal layer. Although the aim of the two was the same, they are two unmixable conceptions with different processing locations of radar and different methods.

6.6 Stable Transition of Equivalent Measurements and Group Termination Algorithm

6.6.1 Algorithm of Stable Transition of Equivalent Measurements

The group scale will change after combination and splitting of group-targets, which leads to the problem of stable transition of equivalent measurement. As for combination, the equivalent measurement in this cycle will not deviate greatly from the predicted center, if it is inside the maximum geometrical contour of the last cycle. Otherwise, the equivalent measurement in this cycle will have big deviation in some direction, if it is outside the maximum geometrical contour and inside target distance of the nearest marginal measurement of the last cycle. As for splitting, marginal targets probably split directly from group-targets, leading to deviation of group-target centroid. Non-marginal targets need some time to become marginal targets first and then split from group-targets. During this time period, group-target centroid slowly changes, and we need to gradually decrease the contribution of splitting targets on equivalent measurement. Group-target centroid deviates even more when faced with batch or module splitting. Thus, the algorithm of stable transition in essence needs to avoid the deviation of equivalent measurement from its tracking gate, when group-target members split and new members merge. Otherwise, tracking will fail. In order not to change the shape of the tracking gate for computation simplicity, we propose the algorithm of stable transition of group-target combination and splitting based on correction of the predicted center.

Group-target member combination will bring new measurements. We only need to assure that equivalent measurement after combination will fall inside tracking gate.

Assume a new measurement which in some direction is outside marginal measurements and inside target distance at time $k - 2$ will be combined to group-target at time k after the validation at time $k - 1$. At time $k - 1$, we predict the center of equivalent measurement without combing the new measurement at time k is as follows:

$$\hat{Y}(k) = H(k)\hat{X}(k/k - 1) \tag{6.37}$$

The combination of new measurement in this cycle k causes the deviation of equivalent measurement at time k. We cannot assure that equivalent measurement falls inside tracking gate, so correct the equivalent measurement at time k as follows:

$$\hat{Y}_c(k) = H(k)\hat{X}(k/k - 1) + [\hat{Y}^e_+ (k) - \hat{Y}^e_- (k)] \tag{6.38}$$

$\hat{Y}^e_+ (k), \hat{Y}^e_- (k)$ are the locations of equivalent measurement at time $k - 1$ with or without the combination of new measurement, respectively. Considering the arrangement of scheduling and that the location changes of members between neighboring cycles will not change greatly, the computation of $\hat{Y}^e_+ (k), \hat{Y}^e_- (k)$ needs to be arranged during validation period at time $k - 1$, and then, the difference between the two variables will be ready before the combination of new members in the next cycle. This can release the resource pressure in the next cycle and assure the instantaneity of the system.

The splitting of group-target members is a process. The measurement that is located on the geometric margin in the last cycle and will immediately split in this cycle will be dealt with similar to new member combination. But the sign problem needs to be solved due to location deviation of equivalent measurement:

$$\hat{Y}_c(k) = H(k)\hat{X}(k/k - 1) - [\hat{Y}^e_- (k) - \hat{Y}^e_+ (k)] \tag{6.39}$$

We discussed the gradual splitting situation in the last section in detail. For target splitting in batch or module, when they are sufficiently far away and dissatisfy the target distance principle, some group-targets will come into being under the condition of association gate forking. At this time, the fork processing of group-target association gates discussed in Chap. 4 can be adopted. Please note, when the number of targets is small, redoing the group-splitting detection can lead to stable transition, which has the merit of simplifying the logic relationships of program structure.

6.6.2 Algorithm of Group Termination

Group termination includes two parts of group-target termination and track termination. Group termination, as the opposite process of group initiation, is also a decision-making method for eliminating the archives of redundant targets, for the sake of avoiding the storage and computation of unnecessary tracks and saving radar system resources. Some references [1] usually discuss the track initiation and termination in multi-target tracking in the same chapter, taking into account the book structure. In fact, track initiation is a process of transforming observation data into tracks, while track termination is a process of judging whether to cancel this track. These two are at the two ends of track processing. Therefore, this book put group initiation and group-splitting detection in one chapter, and put group termination and splitting detection in another chapter. Group-target termination essentially determines whether radar works by means of group-targets. When the number of targets inside group-target association gates is equal to 2, it adopts the independent double-threshold method for splitting centroid detection, terminates group-target tracking, and turn to single-target tracking according to the same principle. Here, we do not give unnecessary details again.

There are many detection methods for single-target track termination such as sequence probability ratio, tracking gate, cost function, and Bayesian method, similar to those for track initiation. This book adopts the method used in group initiation, namely M times of continuous scans and N times of no returns lead to single-target track termination.

6.7 Summary

After analyzing the demerits of the formation target combination and splitting detection methods, we introduce the problem of group-target combination and splitting detection.

As for combination, first, the combination of new observation data and some group-target adopts the association gate principle, regards the observation data that falls inside association gate temporarily as the member of the group-target, and forms a new group-target after validation. This is called association combination method of observation and tracks. Second, the combination of group-targets with stable tracks and/or a single-target group adopts track–track association method to perform combination detection, considering whether track direction and velocity information can support combination. Based on the distributed multi-sensor track fusion algorithm and track history, we propose an independent double-threshold combination algorithm based on track direction and hidden velocity restrictions, namely the double-threshold combination algorithm based on direction velocity restrictions.

After analyzing the influence of location, velocity, and direction of the measurement inside association gates on splitting detection methods, we propose a double association splitting detection algorithm by presetting sub-wave gates within the association gates. (1) Draw lines between the predicted center of equivalent measurement in the last cycle and all measurements inside association gates. (2) Taking the predicted centers of equivalent measurements in this cycle as ends, draw segments that are parallel to and equal in length to the lines drawn between the ends and all measurements in the last cycle. (3) Taking the other end as the predicted center of the measurements in the last cycle, determine the threshold considering the direction information dispersion based on fake directions and velocity information dispersion based on mean displacement (to assure some correct receiving probabilities), and construct sub-wave gates of direction velocity. (4) Taking the sub-wave gates in the last cycle as columns, the measurement in this cycle as rows, and the sub-wave gate as the threshold, construct the association detection matrix of sub-wave gates and measurements, and compute the numbers of the measurements that fall inside and outside sub-wave gates. (5) The measurements falling inside sub-wave gates in this cycle are regarded to have good location-displacement features, while those fallen outside sub-wave gates are possibly false measurements, new measurements, or splitting measurements. We give the latter some "penalty," by assigning a penalty factor (less than 1) according to the deviation of direction and velocity of the measurement from the predicted direction and velocity of equivalent measurement. And then give the equation for computing the penalty factor. (6) Adjust the weights of the measurements outside sub-wave gates according to the penalty factor, and set the equation to compute the equivalent measurements of group-targets under the situation of members' splitting. The method, utilizing the penalty factor to correct the weights of the measurements outside sub-wave gates, avoids drifts with splitting targets and unlimited enlargement of association gates, favors the quick splitting of members from group-targets, watches in real time the moving state of each target (including marginal and inner members of group-targets), and restrains false measurements.

The method presetting sub-wave gates inside association gates is informally similar to the "strainer theory." Association gates are like the body of spoons. Once preset with sub-wave gates, association gates become strainers. Sub-wave gates are the hole in the body of spoons. Measurements are like the "particles" to filter. The detection of the splitting members of group-targets is like filtering these "particles" by using strainers. Put the center of "spoon body" obtained in the last cycle to the predicted center in this cycle, and wait for the new measurements to go through the "hole" of the "spoon body." Leak the measurement with good location-displacement features (again imitating the idea of "mean cinematical behavior" [10] of formation targets), screen out the members about to split or false measurements, and perform next-step processing. The key is where to put the "hole" (the predicted center of sub-wave gates) in the "spoon body" (association gates), how big is the "hole" is (the dimension of sub-wave gates), and how many "holes" (the number of measurement inside association gates in the last cycle) should be put. In terms of the changes of group-targets in scale and shape when some members merge and split, we propose a method that implements stable transition during the procedure of group-target combination and splitting by correcting the predicted center, which assures the stable transition of equivalent measurement when group scale changes. Group termination, as the opposite process of group initiation, is also a decision-making method for eliminating the archives of redundant targets, for the sake of avoiding the storage and computation of unnecessary tracks and saving radar system resources. This book adopts the method used in group initiation, namely M times of continuous scans and N times of no returns lead to single-target track termination.

The simulation results prove the correctness and validity of the group-target combination and splitting detection, and group termination algorithms, as shown in Figs. 7.5, 7.6, 7.17, 7.18–7.21.

Chapter 7
Simulations of Group-Target Tracking Algorithms

7.1 Introduction

We discussed the background, meaning, present situation, content, and idea of group-target tracking algorithms in Chap. 1, some preliminaries in Chap. 2, group splitting detection and group initiation in Chap. 3, single group-target data association and track maintenance in Chap. 4, multi-group-target data association and track maintenance in Chap. 5, and group-target combination and splitting detection and group termination in Chap. 6, respectively. Although the abovementioned algorithm went through strict mathematical reasoning, their correctness and validity still need to be proved by simulations. Simulation validation is parallel to the study of the algorithms. Primary simulation validation are conducted under the ideal condition of no noise waves, after the completion of the group-splitting detection and group initiation algorithm plus the single/multi-group-target data association and track maintenance algorithm. After getting satisfactory results of primary simulation validation, we further study the group-target combination and splitting detection algorithms. Furthermore, the validation of the whole procedure of group-target tracking is conducted after validating all algorithms of group-target tracking.

The system simulation validation is performed on the same simulation platform with same radar parameters and hypotheses. The main focus is to validate the feasibility of the algorithms. Additionally, we only compared the precision of group-target tracking with that of the single-target tracking in the given scenario.

© National Defense Industry Press and Springer Science+Business Media Singapore 2017   143
W. Geng et al., *Group-target Tracking*, DOI 10.1007/978-981-10-1888-6_7

7.2   Simulation Configurations

We design 4 moving targets in a 3-dimensional space. The simulation scenario, parameter design, simulation platform selection, validation items, and standards are as follows:

(1) Simulation scenario

The principles of simulation scenario selection need to cover all the constituents of the algorithms that need to be validated. Thus we design two sparse targets 3 and 4 that move gradually toward targets 1 and 2 with a distance of 10 cm. On the 4th second, the targets 3 and 4 are 20 m away from each other. On the 5th second, they have a distance of 15 m and thus form a group-target. On the 8th second, the targets 1 and 2 are 30 m away from each other. At this moment, the two group-targets' association gates get an intersection. On the 10th second, the target 2 starts to split and thus, we have two split single targets 1 and 2. Then the target 1 keeps its previous flying states.

(2) Parameter design

The track initial value is set to be $X1(0) = [10000, 25, 10100, 12.5, 10000, 0]$, $X2(0) = [10000, 25, 10090, 12.5, 10000, 0]$, $X3(0) = [10000, 25, 9960, 25, 10000, 0]$, $X4(0) = [10000, 25, 9815, 50, 10000, 0]$. The 2-dimensional and 3-dimensional theoretical tracks are shown in Figs. 7.1 and 7.2. The target flies 25 s and the sampling period is $T = 0.1$ s. The radar distance error is $\sigma R = \pm 6$ m, the angle error is $\sigma \theta = \sigma \beta = \pm 0.25$ mrad, and the target model noises are ± 9 m in X, Y, and Z directions.

The velocity in X direction is 25 m/s. The maximum and minimum velocities in Y direction are 50 and 12.5 m/s, respectively. The velocity in Z direction is 0. The simulation data are collected in real time with aforementioned parameter setting. The distance principle of group-target is 20 m, and the group-target's tracking gate and association gate are determined by Eqs. 4.10 and 4.11, respectively.

(3) Simulation platform

We take MATLAB 7.0 as the simulation platform.

(4) Validation items

The validation items include the set of group-target algorithms; group splitting detection and group initiation algorithms; single group-target data association and track maintenance algorithms; multi-group-target data association and track maintenance algorithms; group-target combination; and splitting detection and group termination algorithms. In order to depict the scenario more realistically, we add four uniformly distributed noise waves. The 2-dimensional and 3-dimensional theoretical tracks with both noise waves and process noises are shown in Figs. 7.3 and 7.4.

(5) Validation standards

Given the target distance principles, the group-splitting detection algorithm based on observation data can exactly form group-targets in time, and the group initiation,

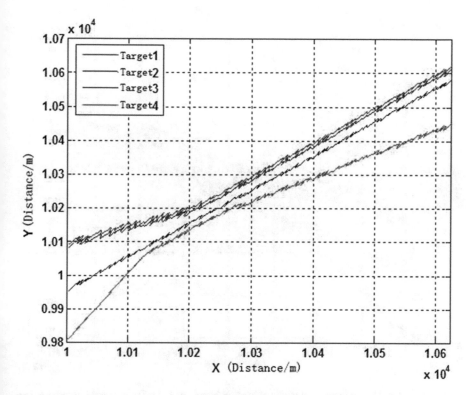

Fig. 7.1 2-dimensional theoretical tracks of the four targets (process noises included)

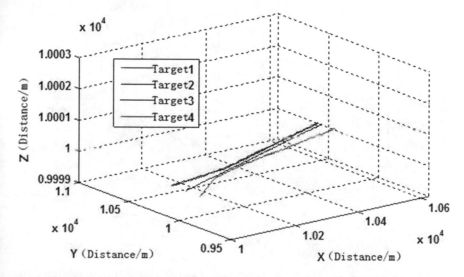

Fig. 7.2 3-dimensional theoretical tracks of the four targets (process noises included)

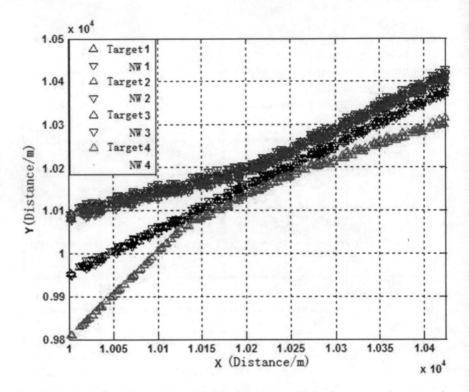

Fig. 7.3 2-dimensional theoretical tracks of the four targets with 4 noise waves and process noises included

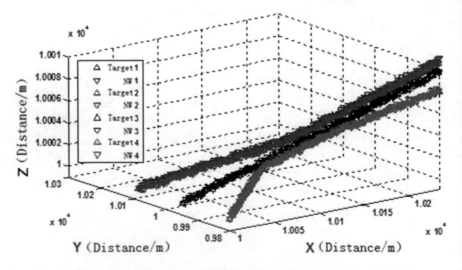

Fig. 7.4 3-dimensional theoretical tracks of the four targets with 4 noise waves and process noises included

based on the measurement of the geometry center of the transient groups, is correct. The single and multi-group-target data association and track maintenance algorithms are correct, with no occurrence of mistaken and missing tracking. The group-target combination algorithms can combine the tracks which satisfy in time the target distance principles. The splitting detection algorithms can correctly separate the split members of the group-target, and can restrain the influence of the noise waves. The group termination algorithms can withdraw group-targets in time and eliminate the unusable tracks. If these algorithms can pass through the validation, the correctness and validity of the group-target tracking algorithms will be proved. In another word, this study can support dense multi-targets tracking.

7.3 Simulation Results and Analyses

7.3.1 The Basic Framework Validation of Group-Targets Tracking

The tracking results of 2-dimensional and 3-dimensional simulations are shown in Figs. 7.5 and 7.6, respectively. First, the group splitting detection module can determine that the targets 1 and 2 satisfy the conditions of group-targets formation, and these two targets get through the group initiation and form a group-target 1. The

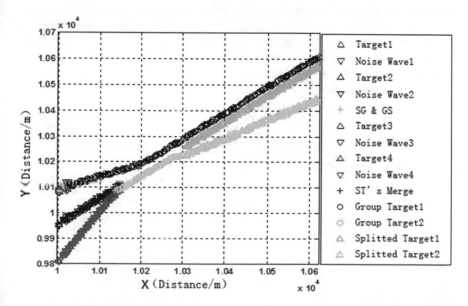

Fig. 7.5 2-dimensional simulation results of the group-targets tracking algorithm considering noise waves and noises

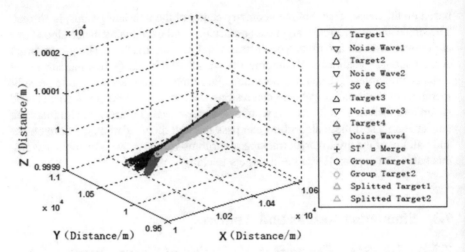

Fig. 7.6 3-dimensional simulation results of the group-targets tracking algorithm considering noise waves and noises

single group-target data association and track maintenance module can relate these two targets and maintain their stable tracks. Second, on the 4th second, the two single targets 3 and 4 merge into a group-target 2 because they two satisfy the targets distance principles and also are tracked normally. Third, on the 5th second, when the association gates of the group-targets 1 and 2 intersect, the simulation goes into the phase of multi-group-target data association and track maintenance, and all the targets are tracked normally. Forth, on the 10th second, when the group-target 2 satisfies the splitting principle, the target 4 splits up and the group-target 2 is eliminated. At this time, targets 1 and 2 come forth. Last, on the 25th second, all tracks are terminated and the tracking ends. In the meantime, the splitting detection module is in charge of noise wave restraining all the time.

7.3.2   Validation of the Group Splitting Detection and Group Initiation Algorithms

The group splitting detection and group initiation are regarded as one of the three key issues, and take an important role in the group-targets tracking algorithms. To validate the group splitting detection and group initiation algorithms and diminish the influence of the complicated track initiation method, we adopt the simple yet common so-called logic initiation—N/M method. The N/M method is used to validate the traditional multi-target tracking with single track initiation and group-targets tracking with group initiation. The simulation results are shown in Figs. 7.7 and 7.8 that magnify some local parts of Fig. 7.5.

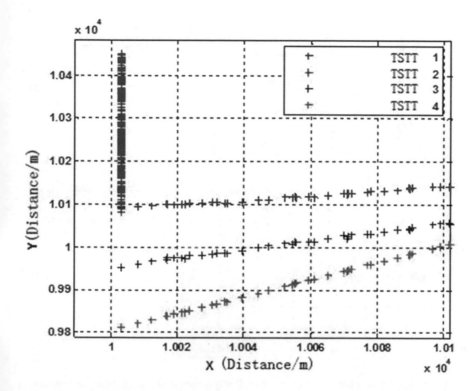

Fig. 7.7 Simulation results of traditional multi-target track initiation considering noise waves and noises

According to the tracking results shown in Fig. 7.7, although the logic track initiation method implements the track initiation of two single targets 3 and 4, it fails in the track initiation of target 2. As shown in Fig. 7.5, two targets 3 and 4 are 145 m away from each other upon the track initiation spot. This distance is far greater than the prescriptive targets distance, so that the track is initiated correctly. However, the distance between dense targets 1 and 2 is only 10 m, which leads to the failure of track initiation with the observation data of a single target.

According to the tracking results shown in Fig. 7.8, the group splitting detection and group initiation algorithm successfully implements not only the group-target formation (based on the targets distance principles), but also the group initiation (based on the measurement of the geometry center of the transient groups). This proves the correctness and validity of the group splitting detection, geometry center computation, and group initiation algorithms investigated in Chap. 3. When considering group-targets tracking algorithm, the group-target 1 and single target 3 have a distance of 135 m upon the track initiation spot (targets 3 and 4 have a distance of 145 m), which presents a three sparse multi-targets (the single targets 3 and 4, and the group-target 1) situation. At this point, the situation naturally leads to

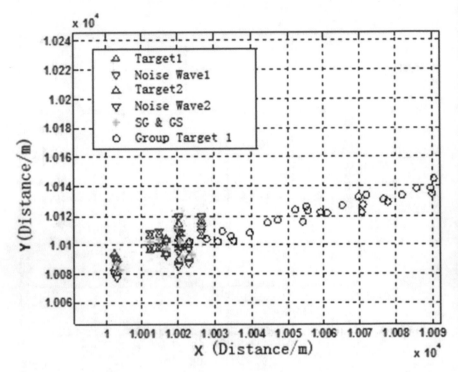

Fig. 7.8 Simulation results of group-targets splitting detection and group initiation considering noise waves and noises

correct initial tracks. This also means that the group-targets tracking algorithm indeed transform the dense multi-targets to sparse multi-targets. The simulation procedure of the group splitting detection and group initiation is shown in Fig. 7.9.

7.3.3 Validation of the Simple Group-Target Data Association and Track Maintenance Algorithm

We use the primary validation under ideal conditions and simulation validation under close-to-realistic environment to validate the algorithm. We consider three targets flying in parallel in a 2-dimensional plane. Only Gauss noises, instead of noise waves, are introduced, and the function of splitting detection is unconsidered. The initial values are $X1(0) = [-22000, 100, 23000, -100]$, $X2(0) = [-22100, 100, 23000, -100]$, $X3(0) = [-22200, 100, 23000, -100]$; timestep $= 120$, $T = 1$ s; the radar distance error is $\sigma R = \pm 20$ m, and angle error is $\sigma \theta = \pm 0.1$ rad. The targets distance principle is 50 m, and thus the association gate is set to be 80 m (fourfold variance). Figures 7.10 and 7.11 show the theoretical track and synthesized track of

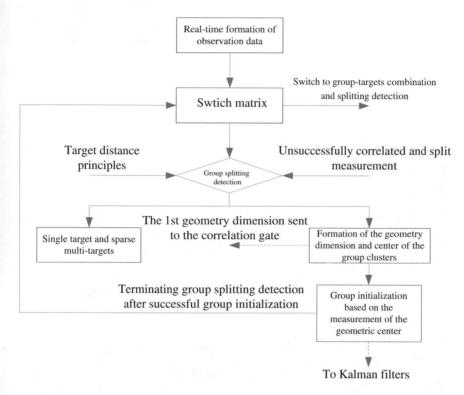

Fig. 7.9 Simulation procedure of the group splitting detection and group initiation

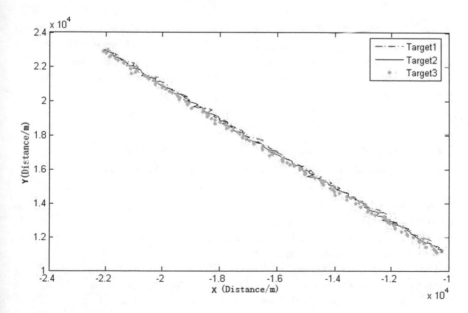

Fig. 7.10 Theoretical tracks of three targets under ideal conditions

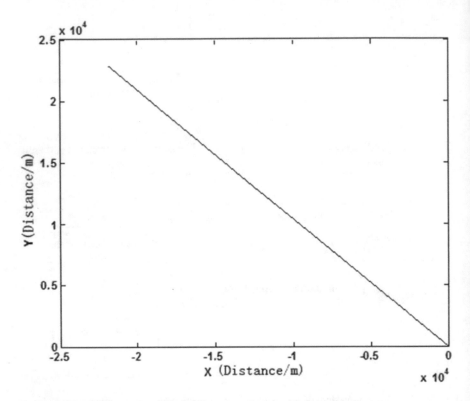

Fig. 7.11 Simulation results of single group-targets under ideal conditions

centroid tracking of the three targets, which validate the algorithm. According to the tracks of the group-targets 1 and 2 given in Fig. 7.5, the tracking is effective. The simulation procedure is shown in Fig. 7.16, excluding the detection matrix of the association gate intersection and the splitting detection matrix of the common measurements.

7.3.4 Validation of the Multi-Group-Target Data Association and Track Maintenance Algorithm

We use also the two-step validation. The first is the simulation validation with no noise waves under ideal conditions, and the second is the system simulation validation with noise waves and splitting detection module in use. MATLAB 7.0 is used as the simulation platform for the both validation.

The simulation scenario under ideal conditions is to consider five targets flying in parallel in a 2-dimensional plane. The targets distance principle is 50 m, and thus, the association gate is set to be 80 m (fourfold variance). The targets 1 and 2

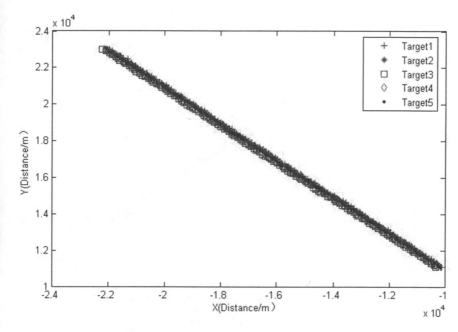

Fig. 7.12   Theoretical tracks of five targets under ideal conditions

and the targets 3 and 4 belong to two different group-targets, respectively. The 5th target is close to the tracks of the targets 1 and 2 and is located within the association gates intersecting area of the two group-targets.

The parameters configurations are as follows: timestep = 120, $T = 1$ s; the distance error is ±20m, the angle error is ±0.1 rad.   The initial values are $X1(0) = [-22000, 100, 23000, -100]$, $X2(0) = [-22050, 100, 23000, -100]$, $X5(0) = [-22200, 100, 23000, -100]$; $X3(0) = [-221500, 100, 23000, -100]$, $X4(0) = [-22200, 100, 23000, -100]$. Figure 7.12 shows the theoretical tracks. Figure 7.13 shows the theoretical tracks with Gauss noises introduced. In Fig. 7.14, the track 1 shows the track of the group-target 1 (including targets 1, 2, and 5), and the track 2 shows the track of the group-target 2 (including targets 3 and 4). The simulation results shown in Fig. 7.14 primarily prove the correctness and validity of multi-group-targets data association algorithm when association gates intersect with no noise waves.

Based on the primary simulation validation under ideal conditions, we then perform the system simulation validation with noise waves and splitting detection module in use.

According to the scenario shown in Fig. 7.1, in the period of 8–10 s, although the targets 2 and 3 have a distance of 30 m, the group-targets 1 and 2 have a distance of 42.5 m (the targets 1 and 2 have a distance of 10 m, the targets 3 and 4 have a distance of 15 m, and the group-targets 1 and 2 should have a distance of 30 m plus 12.5 m, namely 42.5 m). According to Eq. 4.51, the association gates of

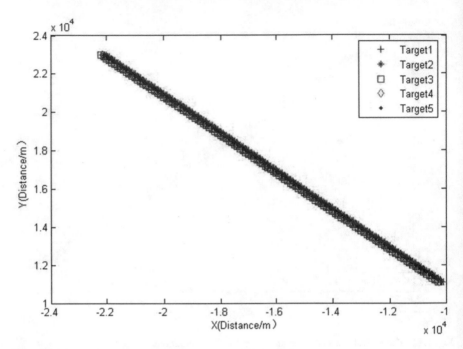

Fig. 7.13 Theoretical tracks of five targets under ideal conditions (Gauss noises introduced)

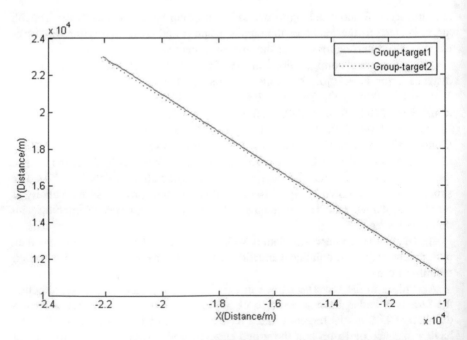

Fig. 7.14 Simulation results of two group-targets tracking when association gates intersect under ideal conditions

the group-targets 1 and 2 are 35 and 37.5 m, respectively. So these two gates intersect but the equivalent measurements of the two group-targets have not entered the association gate of each other. The group-targets combination principle (see Sect. 6.3.2) is "if the distance between the two group-targets is less than the maximum of the two association gate thresholds when the association gates intersect, namely the equivalent measurement of the one is within the association gate of the other, the two group-targets can unite." Obviously, the group-targets 1 and 2 cannot satisfy the principle to combine, so we can only deal with the situation as association gate intersecting of multi-group-targets. Figures 7.5, 7.6 and 7.15 show the tracking results. Figure 7.15 magnifies some local part of 7.5. We can also see that there is no mistaken and missing tracking although the association gates of the two group-targets intersect.

Therefore, both the simulation results under ideal condition and realistic conditions prove the correctness and validity of the multi-group-target data association and track maintenance algorithm under the multi-multicondition. The detailed procedure of multi-group-target data association and track maintenance algorithm is shown as in Fig. 7.16.

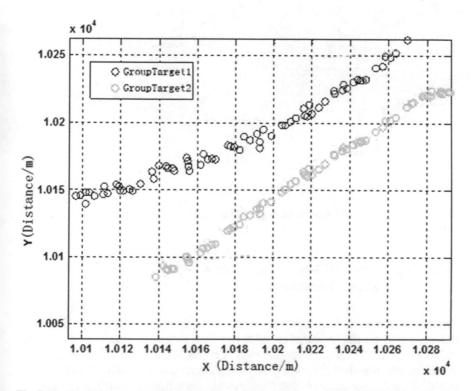

Fig. 7.15 Simulation results of multi-group-targets tracking when association gates intersect under ideal conditions

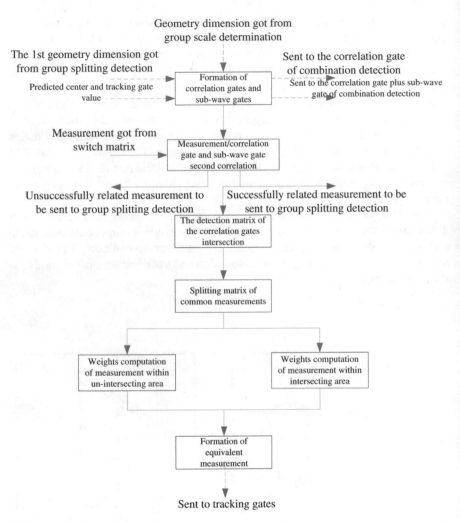

Fig. 7.16 Simulation procedure of multi-group-target tracking algorithm when association gates intersect

7.3.5 Validation of the Group-Target Combination and Stable Transition Algorithm

Combination detection, as a key issue regarding the group-targets tracking algorithm, focuses on the combination of observation data with tracks and the combination of tracks. Because the combination of tracks is much more complicated than that of observation data and tracks, we design the combination of tracks of targets 3 and 4. Once the combination detection module finds that the distance between the targets 3 and 4 satisfies the track distance principle, the two tracks will be combined

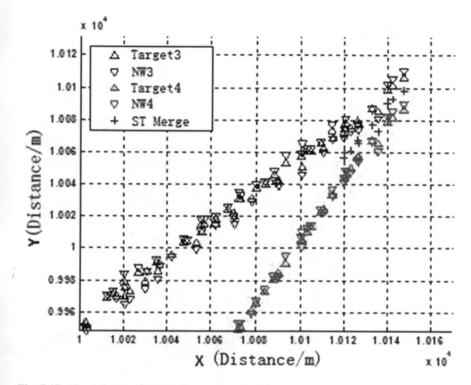

Fig, 7.17 Simulation results of tracks merging algorithms

immediately. The fitting curves in Fig. 7.17 plot both the tracking results and the combination phase of the two tracks, which shows that there is no missing tracking after the combination and the two tracks of the targets 3 and 4 stably change to the track of the group-target 2.

The simulation results verify that the estimation of the direction and velocity before the combination is correct, the independent double-threshold combination algorithm is effective, and the centroid transition is stable.

7.3.6 Validation of the Group-Target Splitting Detection and Group Termination Algorithm

Splitting detection is a very important module of the group-targets tracking algorithm and is a key to the success of group-targets tracking. Since dense moving of multi-targets is a process instead of a purpose, sooner or later they will split. If it can only complete the tracking of group-targets but cannot perform accurate splitting detection, an algorithm fails in its significance and value. So the principle of group-target tracking is to complete single-target tracking whenever needed.

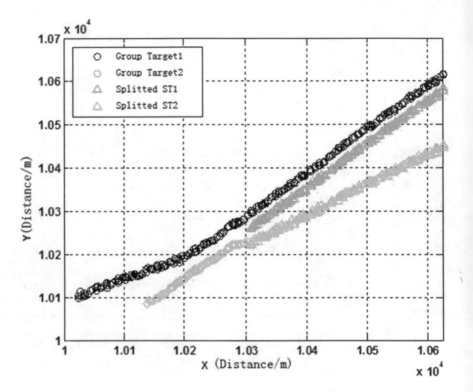

Fig. 7.18 Simulation results of group-targets splitting detection and group termination algorithms

Thus in the subsequent track initiation after the splitting of group-target members, we directly start the track initiation of the split targets without conducting any multi-cycle smoothing. The differences in the track initiation process can be clearly observed in Figs. 7.8, 7.17 and 7.18. Figure 7.18 displays the splitting detection and after-splitting tracking results. In conclusion, (1) the splitting detection module can split in time the targets that dissatisfy targets distance principles; (2) the group termination algorithm can withdraw group-targets correctly; (3) the split single targets can initialize tracks fast (such as the target 2 has been split and initiated quickly); (4) the split target 1 remains stable (under correct tracking) after the splitting of the group-target 2.

The splitting process of the group-target 2 (into targets 1 and 2) and the subsequent tracking process simulate the scenario of what happens after plane missile launching. The group-target tracking is adopted when the plane and the missile cannot be identified from each other. Then the single-target tracking is adopted once the plane and the missile can be split. The simulation results prove the correctness and validity of the splitting detection and group termination algorithm.

7.3.7   Validation of the Dense Target Tracking and False Measurement Restraining Capability

Regarding the splitting detection module of the group-target tracking algorithm, the sub-wave gates preset within the association gates are endowed with noise wave restraining functions, besides the capability of detecting the splitting situation of members. This also needs to be proved in the simulations. Therefore based on the same simulation environment, we adopt the JPDA algorithm to track four targets, and the results are shown in Fig. 7.19. Compared to Figs. 7.5 and 7.19 displays that although one of the targets 1 and 2, together with the targets 3 and 4, was initiated successfully at the beginning, only one track can be given on the 5th second when the targets 3 and 4 have a distance of less than 15 m. This is consistent with the conclusion in Ref. [1], namely that the JPDA algorithm cannot give each target a correct track when targets are too dense. In contrary, the group-targets tracking algorithm initializes the group-target 1 and the targets 3 and 4 and perform the whole-trip tracking of all targets. Therefore, this proves the capability of multi-targets tracking of the algorithm.

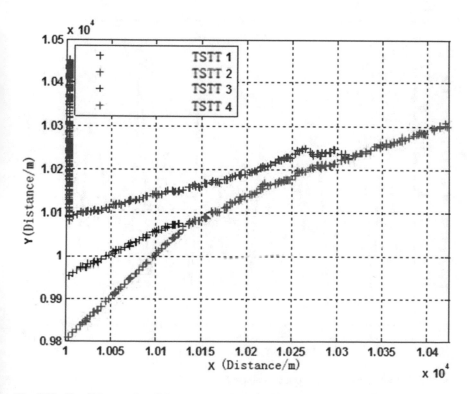

Fig. 7.19 Simulation results of dense group-targets tracking with JPDA algorithm

In fact, the group-targets tracking algorithm has completed the correct tracking of the group-targets 1 and 2, single targets 3 and 4, and split targets 1 and 2 under noise wave conditions. This also proves the noise wave restraining function of the splitting detection module of the algorithm.

To further illustrate the noise wave restraining function of the algorithm, we calculate the absolute precisions of the group-target 2 and the split single target 2 and get 7.4901 and 6.6385 m, respectively. The tracking precision of the group-target 2 is lower than the split target 2 because of two reasons. First, the tracking gate of the group-target 2 is greater than that of the split target 2 (see Sect. 4.3.1 for details on the tracking gate principles of single targets and group-targets). Second, the noise waves in the association gates of the former are more than that of the latter (2 noise waves in the correlation gates of the group-target 2, 1 noise wave in the association gates of the split target 2). This result is satisfying compared to the assumed radar precisions in the simulation design. In conclusion, the splitting detection algorithm can restrain noise waves as expected, and also support stable tracking with given measurement precision. The distributions of the absolute errors are shown in Figs. 7.20 and 7.21.

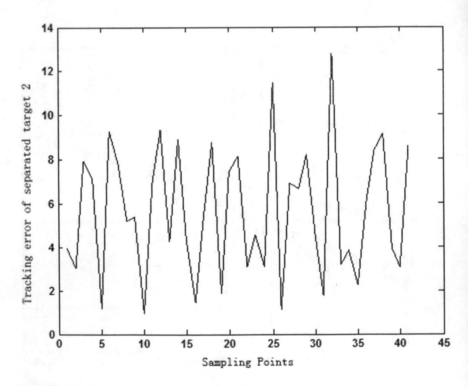

Fig. 7.20 Absolute accuracy of separated target 2

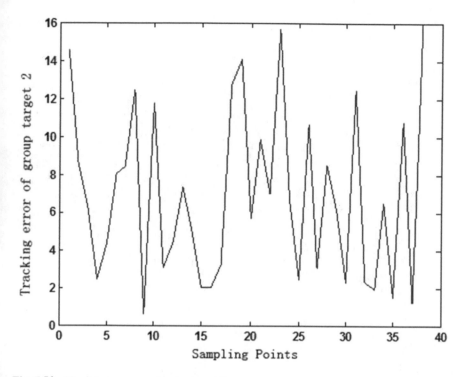

Fig. 7.21 Absolute accuracy of group-target 2

7.4 Summary

This chapter has gone through the simulation validation of all the group-targets tracking algorithms under the conditions of same platform and parameters, etc.

We comprehensively and systematically validate the tracking framework and the algorithms, according to the phases of the simulation procedure planning, scenario, parameter design, simulation platform selection, results analysis, and conclusions. We validate not only the single group-targets tracking algorithm, but also the algorithm regarding the coexistence of multi-group-targets and single targets. The validation content includes the group splitting detection and group initiation, the single group-target data association, the multi-group-target data association, the multi-target and single-target combination, and the group-target splitting detection and group termination, etc.

The simulation results prove the correctness and validity of the algorithms.

References

1. Zhou H, Jing Z, Wang D (1991) Tracking of maneuvering targets. National Defense Industry Press, Beijing
2. Cai Q, Xue Y, Zhang B (1997) Phased array radar data processing and its simulation technology. National Defense Industry Press, Beijing
3. Cai Q, Wang L (1988) Data processing method for tracking of flying targets in formation by phased array radar. Syst Eng Electron 10:38–44
4. Jia P, Wang L (1991) Method for tracking of multi-targets in high density. Mod Defense Technol 2:49–55
5. Sun C, Yuan T (1995) Group tracking of sea targets by 2D search radar. LEIDA YU DUIKANG (4)
6. Wang H, Wang D et al (2001) A new algorithm for group tracking. In: ICR2001, pp 1159–1163
7. Geng W, Liu H et al (2006) A study of Kalman-based algorithm for the maneuvering group target tracking. In: ICR2001, pp 1211–1214
8. Taenzer E (1977) Tracking multiple targets simultaneously with a phased array radar. In: Eascon'77Rec, Wasington, D. C., Sept 1977
9. Taenzer E (1980) Tracking multiple targets simultaneously with a phased array radar. IEEE Trans Aerosp Electron Syst AES-16(5):604–614
10. Binias G (1977) Ulkzielverfolgung (Formation tracking). International reports of the FFM Research Institute (in preparation)
11. Binias G (1978) Basic theory and formation track initiation. International reports of the FFM Research Institute (in preparation)
12. Binias G (1979) The formation tracking procedure for tracking in dense target environment Institute (in preparation)
13. Binias G (1977) Computer controlled racking in dense target environment using a phased array antenna. In: IEE conference publication number 155. Radar 1977, pp 155–159
14. Flad EH (1977) Tracking of formation flying aircraft. In: Rec. IEEE International Radar Conference, London, Oct 1977
15. Shyu HC, Lin YT, Yang JM, Hao JC (1995) The group tacking of targets on sea surface by 2-D search radar. In: Proceedings 1995. IEEE National Radar Conference, 8–11 May 1995, pp 329–333
16. Farina A, Styder FA (1985) Radar data processing, vol I.II. Research Studies Press LTD
17. Tou JT, Gonzalezez RC (1974) Pattern recognition principles. Addison-Wesley Publishing Comp, London
18. Frazier AP, Scott JA (1976) ATOMS-1: an algorithm for tracking of moving sets. System Planning Corporation, Arlington, VA, Report No. ECOM-0510-4,AD-B015080L, Aug 1976
19. Bar-Shalom Y, Jaffer AG (1972) Adaptive nonlinear filtering for tracking with measurement of uncertain. In: Proceedings of 11th conference on decision and control, pp 234–247

20. Hall DL, Llinas J (2001) Handbook of multisensor data fusion. CRC press, Boca Raton
21. Blackman SS (1986) Multiple-target tracking with radar application. Artech House, INC
22. Bar-Shalom Y (1974) Extension of the probabilistic data association filter in multi-target tracking. In: Proceedings of 5th symposium on nonlinear estimation, pp 16–21
23. Bar-Shalom Y (1978) Tracking methods in a multitarget environment. IEEE Trans Autom Control 24(4)
24. Bar-Shalom Y, Campo L (1986) The effect of the common process noise on the two-sensor fused-track covariance. In: IEEE T-AES-22, vol 6, pp 803–805
25. Bar-Shalom Y, Fortmann TE (1988) Tracking and data association. Academic press, New York
26. Bar-Shalom Y (ed) (1990) Multitarget-multisensor tracking: advanced application, vol I. Artech House, INC, Decham, MA
27. Bar-Shalom Y (ed) (1992) Multitarget-mutisensor tracking: advanced application, vol 1. Artech House INC, Decham, MA
28. Bar-Shalom Y, Tse E (1975) Tracking in a cluttered environment with probability data association. Automatica 11(9):451–460
29. Bar-Shalom Y, Li XR (1993) Estimation and tracking: principles, techniques and software. Artech House, Boston
30. Bar-Shalom Y, Li XR (1995) Multitarget-multisensor tracking:principles and techniques. YBS Publishing, Storrs
31. Li XR, Bar-Shalom Y (1996) Multiple-model estimation with variable structure. IEEE Trans Autom Control 41(4):1–16
32. Waltz EL (1986) Data fusion for C3I systems. International C3I handbook. EW communications. Palo Alto
33. Reid DB (1979) An algorithm for tracking multiple targets. IEEE Trans Autom 24(6): 843–854
34. Fortmann TE, Bar-Shalom Y, Scheffe M et al (1981) Detection thresholds for multi-target tracking in clutter. In: Proceedings of the 20th IEEE conference on decision and control, pp 1401–1408
35. Singer RA, Sea RG (1973) New results in optimizing surveillance system tracking and data association performance in dense multitarget environments. IEEE Trans Automat Control AC-18
36. Zhou H (1984) Tracking of maneuvering targets. Ph.D. dissertation, University of Minneasota, Minneapolis
37. Singer RA, Stein JJ (1971) An optimal tracking filter for processing sensor data of imprecisely determined origin in surveillance systems. In: Proceedings of 10th IEEE conference on decision & control, Miami, Beach, FL, pp 171–175, Dec 1971
38. Llinas J, Waltz E (1990) Multisensor data fusion. Artech House, Norwood
39. Hall DL (1992) Mathematical techniques in multisensor data fusion. Artech House, Boston
40. Hall DL, Llinas J (1997) An introduction to multisensor data fusion. Proc IEEE 85(1):6–23
41. Brookner E (1998) Tracking and Kalman filtering made easy. Wiley, New York
42. Maybeck PS (1979) Stochastic models, estimation, and control, vol 1. Academic Press, New York
43. Sorensor W (1985) Kalman Filtering: theory and application. IEEE Press
44. Shafer G (1976) A mathematical theory of evidence. Princeton University Press, Princeton
45. Leduc JP, Mujica F et al (1997) Spatio-temporal wavelet transforms for motion tracking, 0-8186-7919-0/79, IEEE
46. Hong L (1991) Centralized and distributed multisensor integration with uncertainties in communication networks. IEEE Trans AES 27(2):370–379
47. Hong L (1992) Distributed filtering using set models. IEEE Trans AES 28(4):1144–1152
48. Hong L, Lynch A (1993) Recursive temporal-spatial information fusion with applications to target identification. IEEE Trans Aerosp Electron Syst AES-29(2)

49. Hong L (1994) Multirate interacting multiple model filtering for target tracking using multirate models. IEEE Trans Aerosp Electron Syst 30(2):518–524
50. Hong L, Cui NZ, Cong S, Wicher D (1998) An interacting multipattern data association (IMPDA) tracking algorithm. Signal Process 71:55–77
51. Hong L (1999) Multirate interacting multiple model filtering for target tracking using multirate models. IEEE Trans Autom Control 44(7):1326–1340
52. Hong L, Cui N (2000) Interacting multipattern joint probabilistic data association (IMP–JPDA) algorithm for multitargettracking. Signal Process 80(8):1561–1575
53. Hong L, Cui NZ (2001) An interacting multipattern probabilistic data association (IMP–PDA) algorithm for targettracking. IEEE Trans Autom Control 46(8):1113–1238
54. Singer RA, Sea RG (1971) A new filter for optimal in dense multitarget environments. In: Proceedings of ninth Allerton conference circuit and system theory, Urbana Illinois, pp 201–211
55. Yeddanapudi M, BaroShalom Y et al (1995) MATSurv multisensor air traffic surveillance, ADA 292253, USA, Jan 1995
56. Saha RK, Chang KC, Kokar MM (1996) Multisensor track-to-track fusion for airborne surveillance systems. RL-TR-96-123, USA, July 1996
57. Hall D, Nauda A (1989) Embedded AI-based diagnostic system for signal collection and processing systems. J Ware Mater Interact 4(1–3):24–248
58. Drazovich DJ (1983) Sensor fusion in tactical warfare. American Institute of Aeronautics and Antroautics, Inc, pp 1–7
59. Wilson JD, Cantrell BH (1976) Tracking system for asynchronously scanning radars with new association techniques and an adaptive filter. AD-A020540, 14 Jan 1976
60. Cantrell BH, Grindlay A (1980) Multiple site radar tracking system. In: IEEE International Radar Conference
61. Ebert H (1982) Problems of data processing in multiradar and multisensor defense systems. In: IEEE international conference on radar, Oct 1982
62. Waltz EL, Buede DM (1986) Data fusion and decision support for command and control. IEEE Trans SMC 16(16):865–879
63. Hall DL, Llinas J (1987) Data fusion and multisensor correlation. Course notes, Technology Training, Corp, p 327
64. Hall DL, Llinas J (1987) A survey of techniques for CIS data fusion. In: Proceedings of the second international conference on command, control and communications and management information systems. IEEE, Bournemouth, UK, London,pp 77–84
65. Lakin WL, Miles JAH (1985) IKBS in multisensor data fusions. In: Proceedings of IEE conference on advances in C3I, Publication 247, pp 234–240
66. Hall DL, Linn RJ (1990) A Taxonomy of algorithms for multisensor data fusion. In: Proceedings 1990, Tri-service data fusion symposium, pp 13–29, Apr 1991
67. Mabler RPS (1998) Information theoty analysis for data fusion. Lockheed martin tactical defense systems. Eagom, MN
68. Ye X (2003) A study of data association, muld-dimension assignment problem in multitarget tracking. Northwestern Polytechnical University, July 2003
69. Pan Q, Zhang JP, Zhang HC (1994) General probability data association with application to maneuvering multi-target tracking. In: Proceedings of the Asian control conference Tokyo, pp 455–458, 27–30 July 1994
70. Mazor E, Averbuch A, Bar-Shalom Y, Dayan J (1998) Interacting multiple model methods in target tracking: a survey. IEEE Trans AES 34(1):103–123
71. Munir A, Atherton DP (1994) Maneuvering target tracking using an adaptive interacting multiple model algorithm. In: Proceedings of the 1994 American control conference, Baltimore, MD, pp 1324–1328, June 1994
72. Anderson BD, Moore JB (1979) Optimal filtering. Prentice-Hall, New York
73. Maybeck PS (1979) Stochastic models, estimation, and control, vol. 1. Academic Press, New York

74. Stein JJ, Blackman SS (1975) Generalized association of multitarget track data. IEEE Trans Aerosp Electron Syst 11(6)

75. Chang CB, Youens LC (1982) Measurement correlation for multiple sensor tracking in a dense target environment. In: IEEE T-AC-27, vol 6, pp 1250–1252

76. Reid DB (1979) An algorithm for tracking multiple targets. IEEE Trans Autom Control, AC-24:843–854

77. Blom HAP, Bar-Shalom Y (1988) The interacting multiple model algorithm for systems with markovian switching coefficients. IEEE Trans Autom Control 33(8):780–783

78. Formann TE, Bar-Shalom Y, Scheffe M (1983) Sonar tracking of multiple targets using joint probabilistic data association. IEEE J Oceanic Eng 8(3):173–183

79. Chang KC, Bar-Shalom Y (1984) Joint probabilistic data association for multitarget tracking with possibly unresolved measurements and maneuvers. IEEE Trans Autom Control AC-29:585–594

80. Chang KC, Chong CY, Bar-Shalom Y (1986) Joint probabilistic data association in distributed sensor networks. IEEE Trans Autom Control AC-31:889–897

81. Salmond DJ (1990) Mixture reduction algorithms for target tracking in clutter. SPIE Signal Data Process Small Targets 1305:434–445

82. Fitzgerald RJ (1986) Development of practical PDA logic for tracking by microprocessor. In: Proceedings of American control conference, Seattle,WA, pp 889–898, June 1986

83. Roecker JA (1994) A class of near optimal JPDA algorithms. IEEE Trans Aeros Electr Syst 30(2):504–510

84. Zhou B, Bose NK (1993) Multitarget tracking in clutter fast algorithms for data association. IEEE Trans AES-29(2):352–363

85. Fisher JL, Casasent DP (1989) Fast JPDA multitarget tracking algorithm. Appl Opt 28 (2):371–376

86. O'Neil SD, Pao LY (1993) Multisensor fusion algorithms for tracking. In: Proceedings 1993. American control conference. San Francisco, CA, pp 859–863, June 1993

87. Pao LY (1994) Centralized multisensor fusion algorithms for tracking applications. Control Eng Pract 2(5):875–887

88. Deb S, Pattipati KR, Bar-Shalom Y (1992) A multisensor-multitarget data association algorithm for heterogeneous sensors. In: Proceeding of the American and controls conference, pp 1779–1783

89. Deb S et al (1993) A multisensor-multitarget data association algorithm for heterogeneous sensors. IEEE Trans AES 29(2):560–568

90. Deb S, Yeddanapudi M, Pattipati K, Bar-Shalom Y (1997) As generalized S-D assignment algorithm for multisensor-multitarget state estimation. IEEE Trans Aerosp Electr Syst 33 (2):523–537

91. Salmond DJ (1990) Mixture reduction algorithms for target tracking in clutter. SPIE Signal Data Process Small Targets 1305:434–445

92. Watson GA, Blair WD (1992) Tracking maneuvering targets with multiple sensors using the interacting multiple model algorithm. In: Proceedings of signal and data proceeding for small target,SPIE Orlando, FL, Apr 1992

93. Wilson JD (1979) Track initiation techniques in a dense detection environment AD-A07969, Sept 1979

94. Castella FR (1994) Multisensor, multisite tracking filter. IEE Proc Radar Sonar Navig 141 (2):75–82

95. Singer RA, Sea RG, Housewrigth KB (1974) Derivation and evaluation of improved tracking filter for use in dense multitarget environment. IEEE Trans Inf Theory IT-20(4)

96. Kirubarajan T, Bar-Shalom Y, Daeipour E (1995) Adaptive beem pointing control of a phased array radar in the presence of ECM and false alarms using IMMPDAF. In: Proceedings 1995, ACC, Seattle, Washington, pp 2616–2620

97. Chon J, Leung H, Lo T, Litva J, Blanchette M (1996) A modified probabilistic data association filter in a real clutter environment. IEEE Trans AES 32(1):300–313

98. Chang KC, Saha RK, Bar-Shalom Y (1997) On optimal track-to-track fusion. IEEE Trans AES 33(4):1271–1276
99. Bar-Shalom Y (1981) On the track-to-track Association problem. IEEE T-AC-26, vol 2, pp 571–572
100. Bowman CL (1979) Maximum likelihood track association for multisensor integration. IEEE, pp 374–376
101. Bar-Shalom Y (1997) A tutorial on multitarget-multisensor tracking and fusion. In: IEEE national radar conference, Syracuse, NY, 15 May 1997
102. Blair WD, Rice TR, Alouani AT, Xia P (1991) Asynchronous data fusion for target tracking with a multi-tasking radar and optical sensor. SPTE, Acquisition, tracking, and pointing V, vol 1482, pp 234–245
103. Bar-shalom Y, Sherlukde HM, Pattipati KR (1989) Use of measurements from an image sensor for precision target tracking. IEEE Trans AES 25(6):863–871
104. Vidmar A, Malakian K (1994) Optimization of tracking association gate size for maneuvering targets. SPIE, vol 2235, pp 384–387
105. Gauvrit H, Le Cadre IP, Jauffret C (1997) A formulation of multitarget tracking as an incomplete data problem. IEEE Trans AES 33(4):1242–1257
106. Lerro D, Bar-Shalom Y (1993) Tracking with debiased consistent converted measurements versus EKF. IEEE Trans AES 29(4):1015–1022
107. Sun HM, Ching SM (1992) Tracking multitarget in cluttered environment. IEEE Trans AES 28(2):546–559
108. Avityour D (1992) A maximum likelihood approach to data association. IEEE Trans AES 28(2):560–565
109. Fortmann TE, Bar-Shalm Y, Scheffe M (1985) Detection thresholds for tracking in clutter —a connection between estimation and signal processing. IEEE Trans AC 30(3):221–228
110. Li XR, Bar-Shalom Y (1994) Detection threshold selection for tracking performance optimization. IEEE Trans AES 30(3):742–749
111. Ahmeda SS, Harrison I, Woolfson MS (1996) Adaptive probabilistic date-association algorithm for tracking in cluttered environment. IEE Proc Radar Sonar Navig 143(1):17–22
112. Ahmeda SS, Keche M, Harrison I, Woolfson (1997) Adaptive joint probabilistic data association algorithm for tracking multiple targets in cluttered environment. IEE Proc Radar Sonar Navig 144(6):309–314
113. Rong Li X, Bar-shalom Y (1996) Tracking in clutter with nearest neighbour filters analysis and performance. IEEE Trans AES 32(3):995–1009
114. Guu JA, Wei C-H (1991) Tracking technique for maneuvering target with correlated measurement noises and unknown parameters. IEE Proc F 138(3):278–288
115. Rocker JA, Phillis GL (1993) Suboptimal joint probabilistic data association. IEEE Trans AES 29(2):510–517
116. Collins JB, Uhlmann JK (1992) Effective gating in data association with multivariate gaussian distributed states. IEEE Trans AES 28(3):909–916
117. Kim KH (1994) Development of track to track fusion algorithms. In: Proceedings of the American control conference, Maryland, pp 1037–1041, June 1994
118. Rao BS, Durrant-Whyte H (1993) A decentralized Bayesian algorithm for identification of tracked targets. IEEE Trans SMC 23(6):1683–1698
119. Kim H, Swain H (1995) Evidential reasoning approach to multi-source data classification in remote sensing. IEEE Trans SMC 25(8):1257–1265
120. Sadjadi FA (1986) Hypothesis testing in a distributed environment. IEEE Trans Aerosp Electron Syst 22:134–137
121. Waltz EL, Linas J (1990) Multisensor data fusion. Artech House, Norword
122. Wax N (1955) Signal-to-noise improvement and the statistics of tracking populations. J Appl Phys 26:586–595
123. Sittler RW (1964) An optimal data association problem in surveillance theory. IEEE Trans Mil Electr 8(2):125–139

124. He Y, Wang G, Lu D, Peng Y (2010) Multisensor information fusion with applications. Publishing House of Electronics Industry, Beijing
125. He Y, Wang G, Guan X (2010) Information fusion theory with applications. Publishing House of Electronics Industry, Beijing
126. He Y, Xiu J, Guan X (2013) Radar data processing with applications. Publishing House of Electronics Industry, Beijing

Printed in the United States
By Bookmasters